D1160076

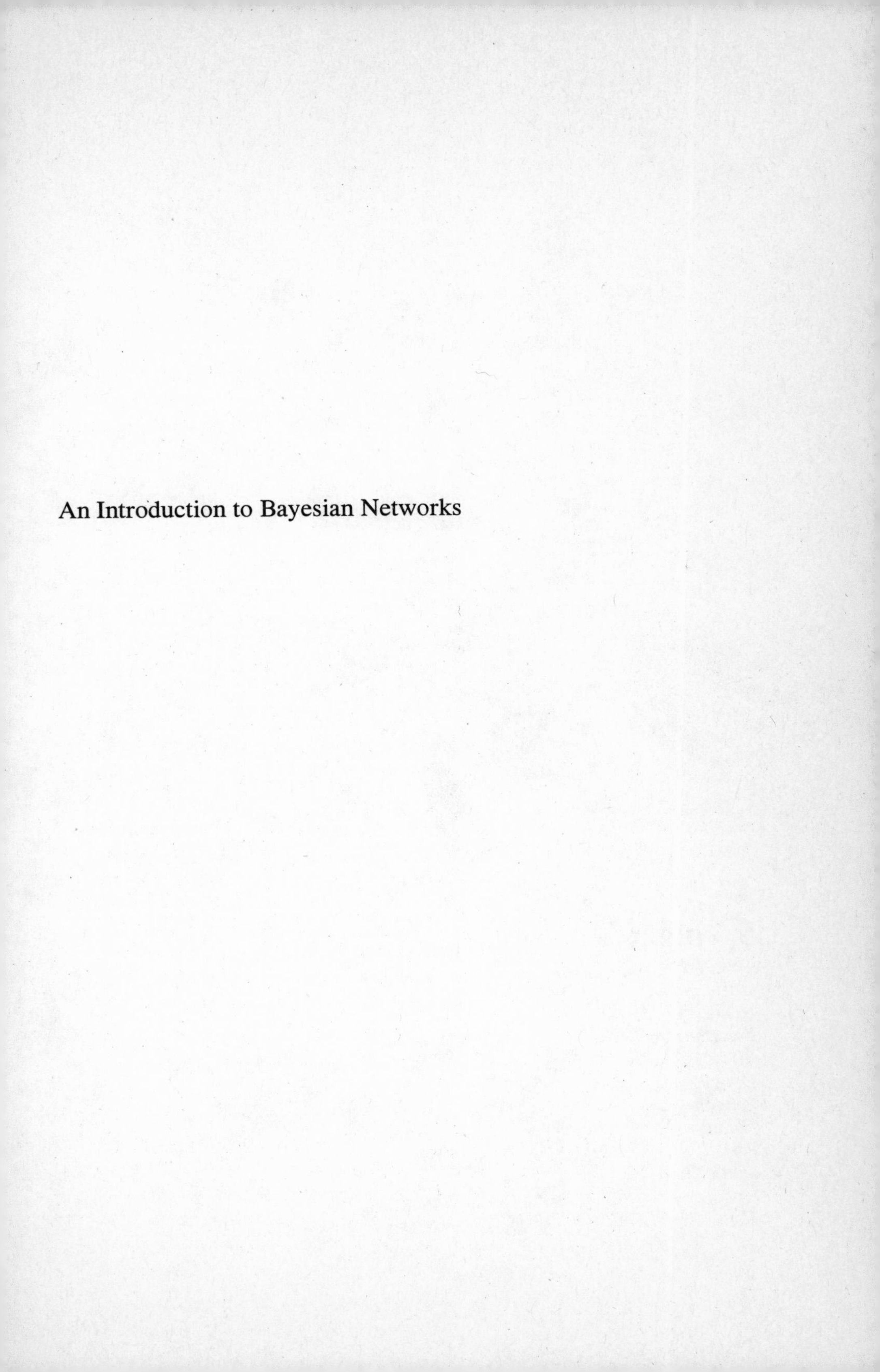

An Introduction to Bayesian Networks

Springer
New York
Berlin
Heidelberg
Barcelona
Budapest
Hong Kong
London
Milan
Paris
Santa Clara
Singapore
Tokyo

An Introduction to Bayesian Networks

Finn V. Jensen
Aalborg University, Denmark

Springer

First published in 1996 by UCL Press
Reprinted 1997

UCL Press Limited
University College London
Gower Street
London WC1E 6BT

The name of University College London (UCL) is a registered trade mark used by UCL Press with the consent of the owner.

ISBN: 0-387-91502-8

Distributed in Canada and the United States
including territories and possessions by

Springer-Verlag New York, Inc.
175 Fifth Avenue
New York, New York 10010

Typeset in Times.
Printed and bound by Bookcraft (Bath) Ltd., England.

Contents

Preface

During the early 1990s Bayesian networks (also called belief networks, Bayesian belief networks or causal probabilistic networks) have attracted a great deal of attention, not only at research institutions but also in industry. Contrary to most other expert system techniques, a good deal of theoretical insight as well as practical experience is required in order to exploit the opportunities provided by Bayesian networks. For newcomers, the field presents itself in a most confusing way. To get a proper insight they have to dig up results in various journal articles and conference proceedings, and there is very little help on practical model building. Also, the literature available is not meant for readers looking for a way into the field.

This book is an attempt to change the present state of affairs. It is intended for both classroom use and self-study, and it addresses persons who are interested in exploiting the Bayesian network approach for the construction of decision support systems or expert systems.

The theoretical exposition in the book is self-contained and the mathematical prerequisite is some prior exposure to elementary calculus, and elementary graph theory. The theory is introduced in an order respecting a practitioner's requirement not to be distracted by theory which does not improve their practice.

Many of the exercises in the book provide practical training; the exercises marked by a ^{H} require a computer system for handling Bayesian networks. For this purpose a PC diskette with a restricted version of HUGIN is included. On the diskette you will find a tool for building Bayesian networks, a library of examples, and also a library of proposed solutions to some of the exercises. (Updated demonstration versions of HUGIN are available through http://www.hugin.dk)

The book falls in two parts. Chapters 1, 2, and 3 provide the background necessary for constructing Bayesian network models. The exposition is organized so that, without relaxing precision, the chapters require only a little mathematical training.

To get beyond the point of mere construction of Bayesian networks and using on-the-shelf software, insight into probability updating is required. Therefore Chapter 4 presents the HUGIN propagation algorithm, and in Chapter 5 the algorithm and modifications of it are used to meet various decision support requirements. Chapter 6 discusses how Bayesian networks can be used when taking decisions on actions. The last three chapters are also self-contained. However, they are mathematically more demanding and require a higher level of mathematical experience from the reader.

As the title indicates, this book is only an introduction to Bayesian networks. Many results on Bayesian networks are not mentioned or only treated superficially. At the end of each chapter you will find references for further reading. These references are in no way complete. The annual *Conference on uncertainty in Artificial Intelligence* is the main forum for researchers working with Bayesian networks. So, the best general reference for further reading on Bayesian networks is the proceedings from these conferences (Morgan Kaufmann Publishers).

The book is organized so that in Chapter 2, Sections 2.1 to 2.3 are the basis for all remaining parts of the chapter, and in Chapter 4, Sections 4.1 to 4.4, are the basis for Chapters 5 and 6. All sections of Chapter 5 can be read independently, and also Chapter 6 is independent of Chapter 5.

A recommended very short course on Bayesian networks would be covered by: Chapter 1 and Sections 2.1 to 2.3, 3.1 and 3.2, 4.1 to 4.4, 5.2 and 5.5, and 6.1 to 6.3.

In this book I have included descriptions of some real world systems. They are only a fraction of the existing systems based on Bayesian networks. At the end of Chapter 1 you will find references to other systems.

I would like to express my gratitude to several people for their help during the preparation of this book. First of all to the ODIN group at Aalborg University (http://www.iesd.auc.dk/odin), in particular to Stig K. Andersen, Frank Jensen, Uffe Kjærulff, Steffen L. Lauritzen, Kristian G. Olesen and Bo Thiesson. This book would never have been produced had it not been for the lively exchange of ideas in the group over the past six years. A special thanks to Uffe Kjærulff who read an early version of the text and provided much constructive criticism and advice. Also thanks to Judea Pearl, David Spiegelhalter and Pedro Larrañaga for their constructive comments and to Søren Lundbye-Christensen for the cartoons in Chapter 1. Finally, I am much indebted to Lisbeth G. Nielsen for her very competent LaTeX-coding and for several corrections to my English language.

Finn V. Jensen
Aalborg, October 1995

Chapter 1

Introduction

In Figure 1.1 you see an expert at work. Her domain of expertise is a well defined part of the world. She may be a physician examining a patient, she may be a banker questioning an applicant for a loan, or she may be a pilot monitoring the aircraft.

Figure 1.1 An expert at work.

A simple view of the tasks of the expert is the triangle in Figure 1.2. First of all she observes her part of the world to establish the state of it. If she is a physician she will use the patient's reported symptoms and his medical record to come up with her diagnosis. The banker will use information on the applicant's finances to evaluate the chance for repayment, and the pilot will watch the instruments in the cockpit to (hopefully) make sure that everything is OK.

Based on her interpretation of the state of the world, the expert decides on an action. An action is an intervention in her part of the world. It can be the treatment of a patient, settling the terms of a loan, or deciding not to make any changes.

For any action the expert has some expectations. Sometimes they come true and sometimes they do not; but in any case she will learn from the results of the actions which may help her in interpreting the world in future.

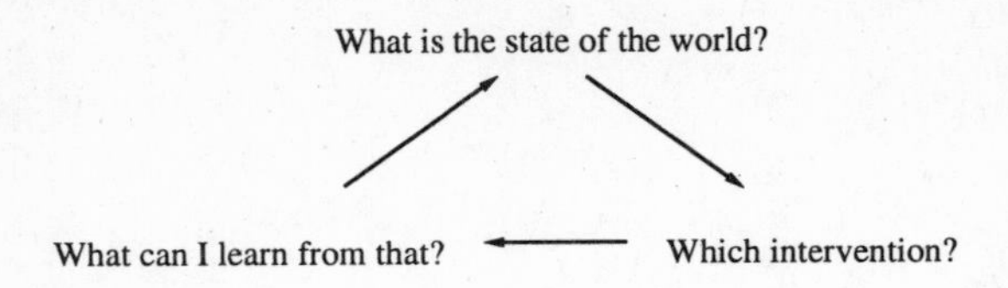

Figure 1.2 The task-triangle of an expert.

1.1 Rule based systems

The first *expert systems* were constructed in the late 1960s. Their scope is decision making characterized by repeatedly deciding on almost similar cases. The systems are constructed as computer models of the expert (see Fig. 1.3).

Figure 1.3 An expert system. The expert is modelled by the computer.

The vision of the technology was that experts could be replaced by computer systems which modelled the best experts in the world. The building blocks for modelling the expert were *production rules.* A production rule is of the form

if *condition* **then** *fact*

or

if *condition* **then** *action,*

where the condition is a logical expression. A so-called *rule based system* consists of a *knowledge base* and an *inference system.* The knowledge base is a set of production rules and the inference system combines rules and observations to come up with conclusions on the state of the world and on actions. Though the language is very simple, it turned out to be rather powerful when modelling experts' reasoning, and several impressive rule based expert systems were constructed (for example MYCIN (Shortliffe 1976) and R1 (McDermott 1984)).

1.2 Uncertainty

Rather soon after their first successes it became clear that rule based systems have their shortcomings. One of the major problems was how to treat uncertainty. In most expert domains you will meet a need for reasoning under uncertainty. There are various sources of uncertainty. The observations may be uncertain, the information may be incomplete, the relations in the domain may be of a non-deterministic type (like diseases and symptoms). It may be that you, in your reasoning, abstract from less important factors, or it may be that the terms involved are vague (like "large", "beautiful", or "pain").

A way to incorporate uncertainty in rule based systems is to extend the production rules to the format

if *condition* with certainty x **then** *fact* with certainty $f(x)$,

where f is a function.

The inference system must be extended with new *inference rules*, which shall ensure a coherent reasoning under uncertainty. For example, if we can infer C from A with certainty x, and we can infer C from B with certainty y, what can we conclude on the certainty of C? The inference rule specifies a function $g(x, y)$.

However, it is not possible to capture reasoning under uncertainty with inference rules for production rules. The reason being is that inference rules are context free while coherent reasoning under uncertainty is sensitive to the context in which the certainties have been established. We shall deal with this problem more precisely later.

1.3 Normative expert systems

In the field of *decision theory*, classical probability theory has been extended to a very precise mathematical framework for rational decision making, and experience shows that experts generally do not obey the rules of this framework. So, it is argued that experts are not good at quantitative reasoning under uncertainty, and therefore they need support from the computer (see Fig. 1.4).

Normative expert systems are an alternative to rule based expert systems. Both types of system deal with repeated decision making on almost similar cases, but the design principles for normative systems differ from rule based systems in three ways:

- instead of modelling the expert, model the domain;

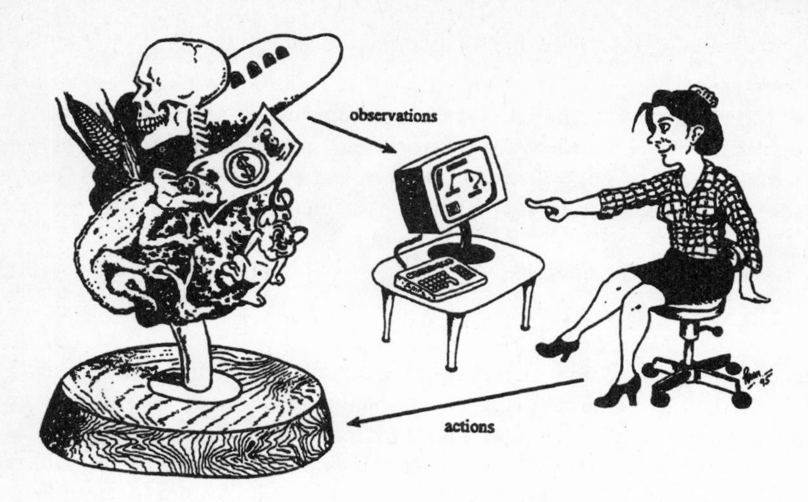

Figure 1.4 A normative expert system. The domain is modelled, and the computer gives advice.

- instead of using a noncoherent uncertainty calculus tailored for rules, use classical probability calculus and decision theory;
- instead of replacing the expert, support her.

Historically, the principles above are not new. In the 1960s attempts were already being made to use classical probability theory in expert systems (Gorry & Barnett 1968). However, due to the very heavy calculation load required the attempt was abandoned and considered an intractable task (Gorry 1973).

In the mid 1980s the principles were revived. Work by Pearl (1986b) introduced Bayesian networks to expert systems, and with the MUNIN system (Andreassen et al. 1989) it was shown that the necessary calculations for very large networks are indeed tractable. This – and what happened since – is what this book is about.

1.4 Some references for normative systems

Below we have listed references for normative systems. The list is not an attempt to be complete, it should merely give an impression of the variety of domains.

Agriculture

BOBLO is a system which helps in verification of the parentage of Jersey cattle through blood type identification (Rasmussen 1995a,b). The system is described in Section 2.5.

A system for mildew control in winter wheat has been constructed by Jensen (1995a). In Chapter 6 we give simplified examples from this system.

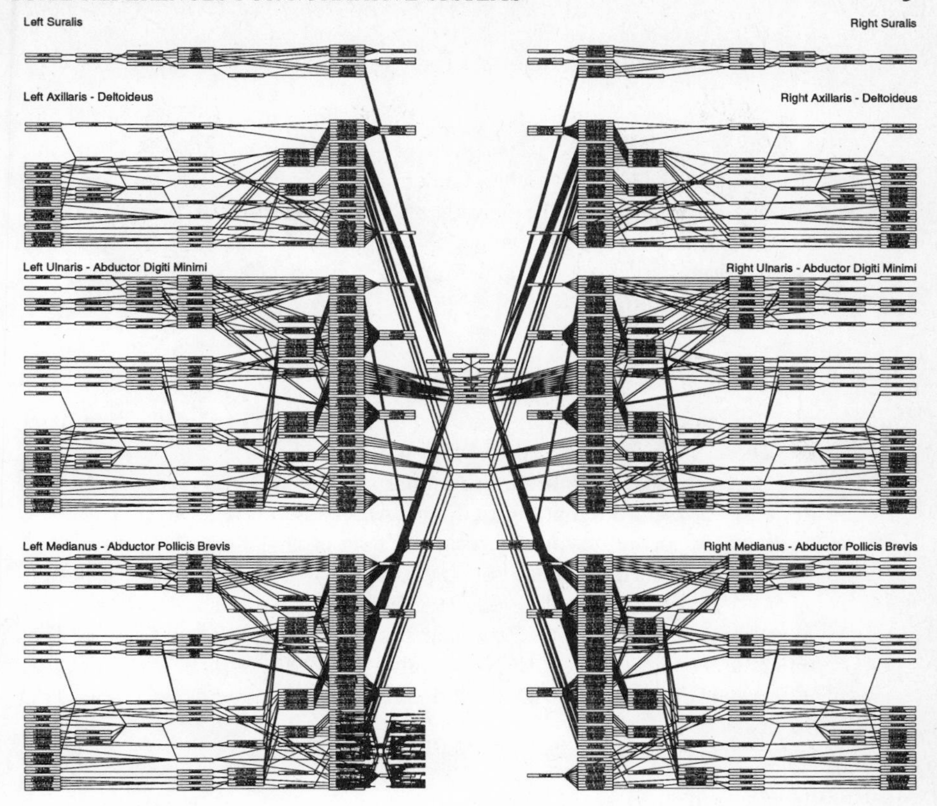

Figure 1.5 MUNIN is a system for diagnosing diseases in muscles and nerves. The network above is the model for six muscles and eight nerves. It contains a little more than 1000 nodes. Updating time is approximately ten seconds on a 1990 SUN work station.

Computer vision

Binford and colleagues made an early use of Bayesian networks for interpretation of images (Binford et al. 1988, Levitt et al. 1989). Jensen et al. (1992) and Rimey & Brown (1994) use Bayesian networks for control of computer resources in the interpretation process, and Munck-Fairwood (1992) uses a Bayesian network for 3D inference from 2D data. The Bayesian network approach to image interpretation has been used by Levitt et al. (1993) for segmentation of computed radiographs of the hand.

Computing

The PC-operating system *Windows 95* incorporates a normative system for trouble-shooting printing problems. The principles behind it are described in Heckerman et al. (1995).

DAACS is a system for software debugging (Burnell & Horvitz 1995).

Information processing

VISTA is a system used by NASA when launching space shuttles. Its purpose is to filter and display information on the propulsion system (Horvitz & Barry 1995).

Bruza & van der Gaag (1993) developed a language for constructing Bayesian networks for information retrieval, and Fung & Favero (1995) describe another system for information retrieval.

Medicine

Child helps in diagnosing congenital heart diseases (Franklin et al. 1989, Lauritzen et al. 1994). The system is described in Section 3.5.

MUNIN is a system for obtaining a preliminary diagnosis of neuromuscular diseases on the basis of electromyografic findings (Andreassen et al. 1989).

Painulim diagnoses neuromuscular diseases (Xiang et al. 1993).

Pathfinder is of assistance to community pathologists with the diagnosis of lymph-node pathology (Heckerman et al. 1992, Heckerman & Nathwani 1992a,b). The system is described in Section 5.6. Pathfinder has been integrated with videodiscs to the commercial system *Intellipath* (Nathwani et al. 1990).

SWAN is a system for insulin dose adjustment of diabetes patients (Andreassen et al. 1991, Hejlesen et al. 1993).

Miscellaneous

Hailfinder was developed for forecasting severe weather in the plane of northeastern Colorado (Abramson et al. 1996).

FRAIL is an automatic Bayesian network construction system (Goldman & Charniak 1993). It has been developed for building Bayesian networks for interpretation of written prose (Charniak & Goldman 1991).

Chapter 2

Causal and Bayesian networks

This chapter introduces *causal networks* as graphical representations of causal relations in a domain. Through several examples, basic rules for chained reasoning about certainty are introduced. These rules are formalized in the concept of *d-separation.*

In Section 2.3 we present the probability calculus used in this book, and we define the concept of a *Bayesian network.* In Section 2.4 the introductory examples are modelled as Bayesian networks and the reasoning is performed through probability calculations.

Finally we describe the BOBLO system.

2.1 Examples

In this section we give three examples. They illustrate crucial points to consider when reasoning about certainty has to be formalized.

2.1.1 Icy roads

Police Inspector Smith is impatiently awaiting the arrival of Mr Holmes and Dr Watson; they are late and Inspector Smith has another important appointment (lunch). Looking out of the window he wonders whether the roads are icy. Both are notoriously bad drivers, so if the roads are icy they may crash.

His secretary enters and tells him that Dr Watson has had a car accident. "Watson? OK. It could be worse ... icy roads! Then Holmes has most probably crashed too. I'll go for lunch now."

"Icy roads?", the secretary replies, "It is far from being that cold, and furthermore all the roads are salted." Inspector Smith is relieved. "Bad luck for Watson. Let us give Holmes ten minutes more."

To formalize the story, let the events be represented by variables with two states, *yes* and *no*. Suppose also that to each event is associated a *certainty*, which is a real number. So, we have the three variables: *icy roads (I), Holmes crashes (H)* and *Watson crashes (W). I* has the effect of increasing the certainty of both H and

W. We may think of the impact as an increasing function from the certainty of the cause to the certainty of the effect. The situation is illustrated in Figure 2.1.

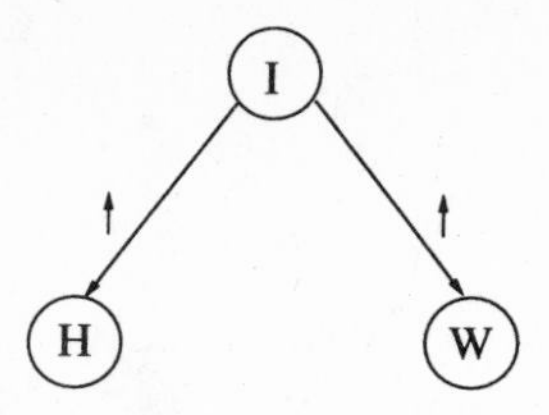

Figure 2.1 A network model of *icy roads*. The arrows on the links model the causal impact, and the small arrows attached to the links indicate the direction of the impact on the certainty.

When Inspector Smith is told that Watson has had a car accident, he is doing a reasoning in the opposite direction to the causal arrows. Since the impact function pointing at W is increasing, the inverse function is also increasing. Hence, he gets an increased certainty of I. The increased certainty of I in turn creates a new expectation, namely an increased certainty of H.

Next, when his secretary tells him that the roads cannot possibly be icy, the fact that Watson has crashed cannot change his expectation concerning road conditions and, consequently, Watson's crash has no influence on H.

This is an example of how dependence/independence changes with the information at hand. When nothing is known about the condition of the roads, then H and W are *dependent*: information on either event affects the certainty of the other. However, when the condition of the roads is known for certain, then they are *independent*: information on W has no effect on the certainty of H and vice versa. This phenomenon is called *conditional independence.*

2.1.2 Wet grass

Mr Holmes now lives in Los Angeles. One morning when Holmes leaves his house, he realizes that his grass is wet. Is it due to rain (R), or has he forgotten to turn off the sprinkler (S)? His belief in both events increases.

Next he notices that the grass of his neighbour, Dr Watson, is also wet. Elementary: Holmes is almost certain that it has been raining.

A formalization of the situation is shown in Figure 2.2.

When Holmes notices his own wet grass, he is doing a reasoning in the opposite direction to the causal arrows. Since both impact functions pointing at H are increasing, his certainty of both R and S increases. The increased certainty of R in turn creates an increased certainty of W.

Therefore Holmes checks Watson's grass, and when he discovers that it is also wet, he immediately increases the certainty of R drastically.

The next step in the reasoning is hard for machines, but natural for human beings, namely *explaining away*: Holmes' wet grass has been explained and thus there is

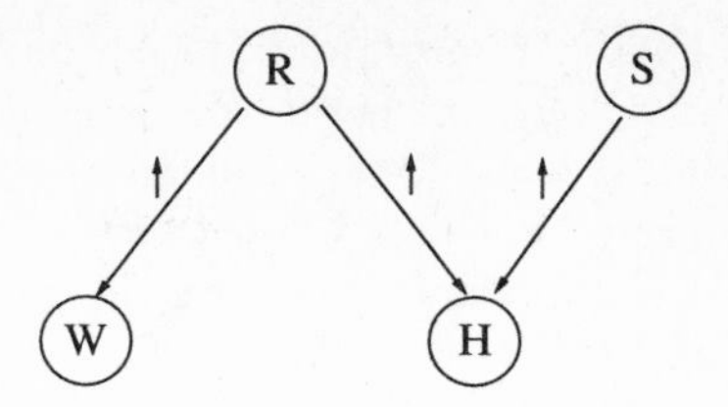

Figure 2.2 A network model for the *wet grass* example. *Rain* and *sprinkler* are causes of *Holmes'* grass being wet. Only *rain* can cause *Watson's* grass to be wet.

no longer any reason to believe that the sprinkler has been on. Hence, the certainty of S is reduced to its initial size.

Explaining away is another example of dependence changing with the information available. In the initial state, when nothing is known, R and S are independent. However, when we have information on Holmes' grass, then R and S become dependent.

2.1.3 Causation and reasoning

A possible source of confusion should be sorted out at this point. The graphs in Figures 2.1 and 2.2 were presented as models for impacts between events, but the reasoning based on the graphs is concerned with how our certainty of the various events is affected by new certainty of other events.

Actually, the models are guidelines for ways of reasoning about unknown events. When reasoning in the direction of the links, the statement in the model is:

The event A causes with certainty x the event B.

From this we reason:

If we know that A has taken place, then B has taken place with certainty x.

Reasoning in the opposite direction to the links is more delicate. So far we have only said that the certainty of the cause A increases when the consequence B has taken place. If you want to get a quantitative statement, your certainty calculus must have a way of inverting the causal statements. In Section 2.4 we show that for probability calculus, Bayes' rule is used for the inversion.

Some scientists take the point of view that the networks are not causal models, but models for how information may propagate between events. This is, from a foundational point of view, perfectly valid as long as you do not model interfering actions in your network. We shall expand on this in Chapter 6.

2.1.4 Earthquake or burglary

Mr Holmes is working at his office when he receives a telephone call from Watson, who tells him that Holmes' burglar alarm (A) has gone off. Convinced that a burglar

(B) has broken into his house, Holmes rushes to his car and heads for home. On his way he listens to the radio (R), and in the news it is reported that there has been a small earthquake (E) in the area. Knowing that earthquakes have a tendency to turn the burglar alarm on, he returns to his work leaving his neighbours the pleasure of the noise. Figure 2.3 gives a model for the reasoning.

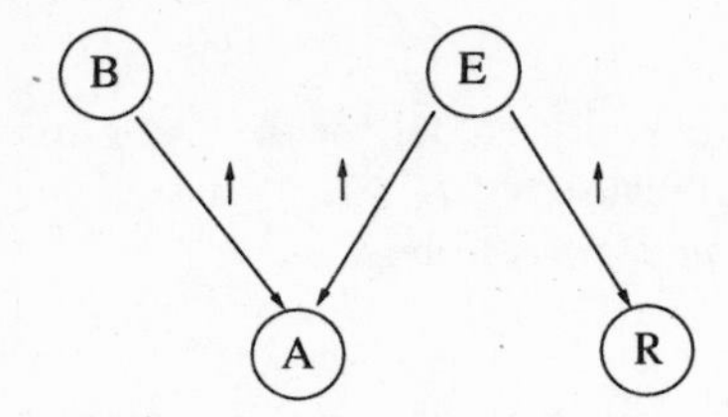

Figure 2.3 A model for the *earthquake* example. Notice that the structure is similar to Figure 2.2.

2.1.5 Prior certainties

It has been typical of the reasoning in the examples of this section that if some event is known, then the certainty of other events must be changed. If, in a certainty calculus, the actual certainty of a specific event has to be calculated, then knowledge of certainties prior to any information is also needed. In particular, prior certainties are required for the events which are not effects of causes in the network.

Take for instance the *wet grass* example. Given that Holmes' grass is wet, the certainty of R is still dependent on whether rain at night is a rare event (as in Los Angeles) or very common (as in London).

The same goes for the earthquake in Section 2.1.4. Though E may have a stronger effect on A than B has, and therefore information on A will increase the certainty of earthquake more than on burglary, the resulting certainty on E should still be lower than the certainty on B. To be able to do this reasoning, prior certainties on E and B are required.

2.2 Causal networks and d-separation

The models in Section 2.1 are examples of *causal networks*. A causal network consists of a set of *variables* and a set of *directed links* between variables. Mathematically the structure is called a directed graph. When talking about the relations in a directed graph we use the wording of family relations: if there is a link from A to B we say that B is a *child* of A, and A is a *parent* of B.

The variables represent events (propositions). In Section 2.1, each variable had the states *yes* and *no* reflecting whether a certain event had taken place or not. In general, a variable can have any number of states. A variable may, for example, be the colour of a car (states $blue, green, red, brown$), the number of children in a family (states $0, 1, 2, 3, 4, 5, 6, > 6$), or a disease (states *bronchitis, tuberculosis, lung cancer*). Variables may have a countable or a continuous state-set, but in this book we solely consider variables with a finite number of states.

In a causal network a variable represents a set of possible states of affairs. A variable is in exactly one of its states; which one may be unknown to us.

Reasoning about uncertainty also has a quantitative part, namely calculation and combination of certainty numbers. The considerations in this section are independent of the particular uncertainty calculus. Whatever calculus is used, it must obey the rules illustrated in Section 2.1 that we formalize in this section.

Serial connections

Consider the situation in Figure 2.4. A has an influence on B which in turn has influence on C. Obviously, evidence on A will influence the certainty of B which then influences the certainty of C. Similarly, evidence on C will influence the certainty on A through B. On the other hand, if the state of B is known, then the channel is blocked, and A and C become independent. We say that A and C are *d-separated given* B, and when the state of a variable is known we say that it is *instantiated.*

We conclude that evidence may be transmitted through a serial connection unless the state of the variable in the connection is known.

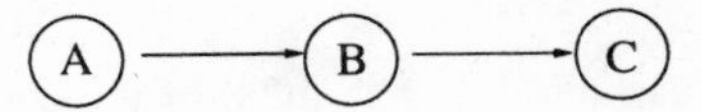

Figure 2.4 Serial connection. When B is instantiated it blocks communication between A and C.

Diverging connections

The situation in Figure 2.5 is a generalization of the *icy roads* example. Influence can pass between all the children of A unless the state of A is known. We say that $B, C, \ldots, E$ are d-separated given A.

So, evidence may be transmitted through a diverging connection unless it is instantiated.

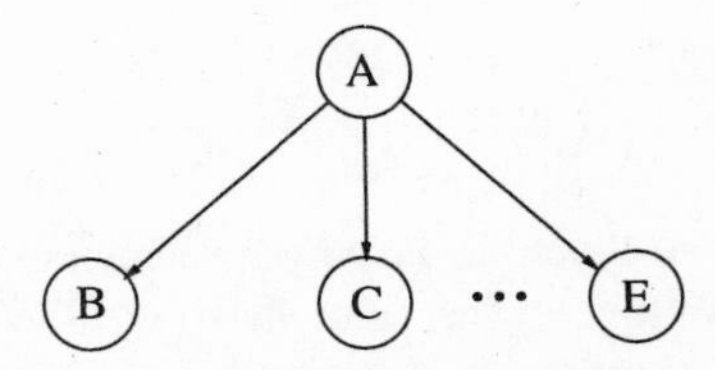

Figure 2.5 Diverging connection. If A is instantiated, it blocks communication between its children.

Converging connections

A description of the situation in Figure 2.6 requires a little more care. If nothing is known about A except what may be inferred from knowledge of its parents $B, \ldots, E$, then the parents are independent: evidence on one of them has no influence on the certainty of the others.

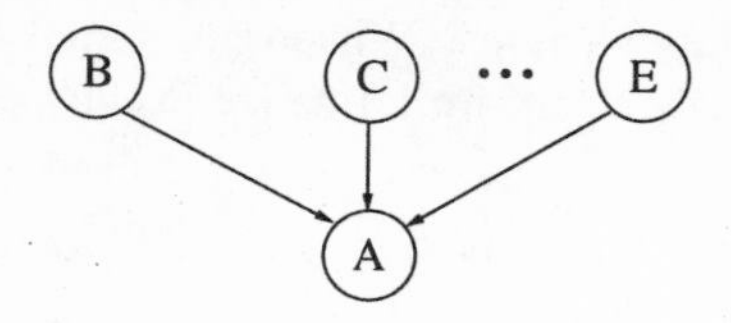

Figure 2.6 Converging connection. If A changes certainty, it opens communication between its parents.

Now, if any other kind of evidence influences the certainty of A, then the parents become dependent due to the principle of explaining away. The evidence may be direct evidence on A, or it may be evidence from a child. This phenomenon is called *conditional dependence*. In Figure 2.7 some illustrating examples are listed.

The conclusion is that evidence may only be transmitted through a converging connection if either the variable in the connection or one of its descendants has received evidence.

Remark. Evidence on a variable is a statement of the certainties of its states. If the statement gives the exact state of the variable we call it *hard* evidence, otherwise it is called *soft*. Hard evidence is also called *instantiation*. Blocking in the case of serial and diverging connections requires hard evidence, while opening in the case of converging connections holds for all kinds of evidence.

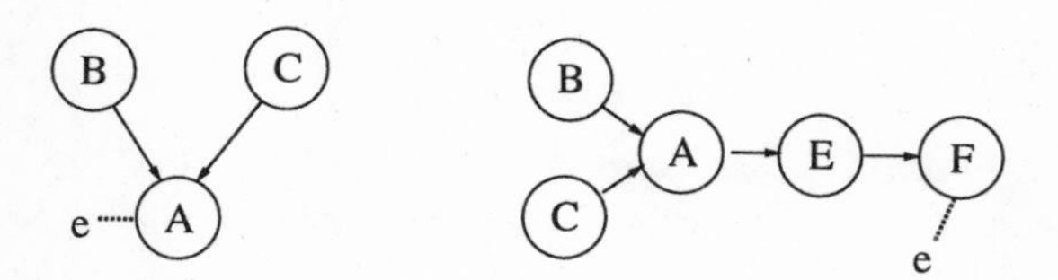

Figure 2.7 Examples where the parents of A are dependent. The dotted lines indicate insertion of evidence.

2.2.1 d-separation

The three cases given above cover all the ways in which evidence may be transmitted through a variable, and following the rules it is possible to decide for any pair of variables in a causal network whether they are dependent given the evidence entered into the network. The rules are formulated in the following.

Definition (d-separation). Two variables A and B in a causal network are *d-separated* if for all paths between A and B there is an intermediate variable V such that either

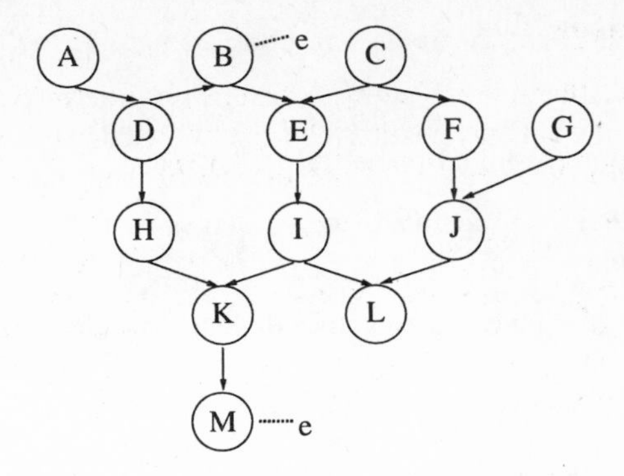

Figure 2.8 A causal network with B and M instantiated. A is d-separated from G only.

– the connection is serial or diverging and the state of V is known

or

– the connection is converging and neither V nor any of Vs descendants have received evidence.

If A and B are not d-separated we call them *d-connected.*

Figure 2.8 gives an example of a larger network. The evidence entered at B and M represents instantiation. If evidence is entered at A it may be transmitted to D. The variable B is blocked, so the evidence cannot pass through B to E. However, it may be passed to H and K. Since the child M of K has received evidence, evidence from H may pass to I and further to E, C, F, J and L. So, the path $A - D - H - K - I - E - C - F - J - L$ is a d-connecting path.

Figure 2.9 gives two illustrating examples.

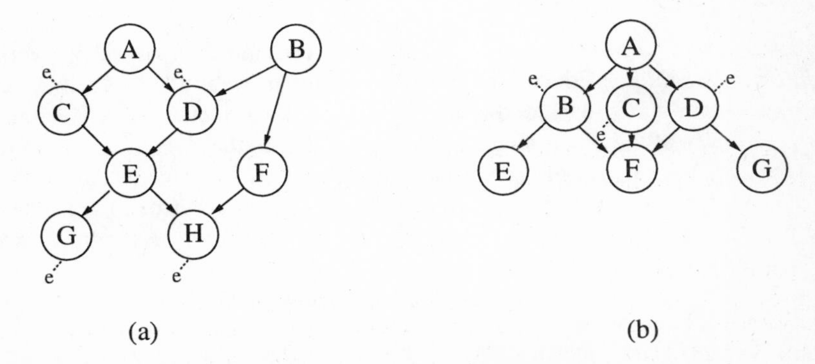

Figure 2.9 Causal networks with hard evidence entered (the variables are instantiated). (a) Although all neighbours of E are instantiated it is d-connected to F, B and A. (b) F is d-separated from the remaining un-instantiated variables.

Note that although A and B are d-connected, changes in the belief in A need not change the belief in B.

You may wonder why we have introduced d-separation as a definition rather than as a theorem. A theorem should be as follows.

Claim. If A and B are d-separated, then changes in the certainty of A have no impact on the certainty on B.

However, the claim cannot be established as a theorem without a more precise description of the concept of "certainty". You can take d-separation as a property of human reasoning and require that any certainty calculus must comply with the claim.

2.3 Bayesian networks

So far nothing has been said about the quantitative part of certainty assessment. Various certainty calculi exist, but in this book we only treat the so called Bayesian calculus, which is *classical probability calculus*.

2.3.1 Basic axioms

The probability $P(A)$ of an event A is a number in the unit interval [0, 1]. Probabilities obey the following basic axioms.

(i) $P(A) = 1$ if and only if A is certain.

(ii) If A and B are mutually exclusive, then

$$P(A \vee B) = P(A) + P(B).$$

2.3.2 Conditional probabilities

The basic concept in the Bayesian treatment of certainties in causal networks is *conditional probability*. Whenever a statement of the probability, $P(A)$, of an event A is given, then it is given conditioned by other known factors. A statement like "The probability of the die turning up 6 is $\frac{1}{6}$" usually has the unsaid prerequisite that it is a fair die – or rather, as long as I know nothing of it, I assume it to be a fair die. This means that the statement should be "Given that it is a fair die, the probability ...". In this way, any statement on probabilities is a statement conditioned on what else is known.

A conditional probability statement is of the following kind:

Given the event B, the probability of the event A is x.

The notation for the statement above is $P(A \mid B) = x$.

It should be stressed that $P(A \mid B) = x$ does not mean that whenever B is true then the probability for A is x. It means that if B is true, and *everything else known is irrelevant for A*, then $P(A) = x$.

The *fundamental rule* for probability calculus is the following:

$$P(A \mid B)P(B) = P(A, B), \tag{2.1}$$

where $P(A, B)$ is the probability of the joint event $A \wedge B$. Remembering that probabilities should always be conditioned by a context C, the formula should read

$$P(A \mid B, C)P(B \mid C) = P(A, B \mid C). \tag{2.2}$$

From 2.1 it follows that $P(A \mid B)P(B) = P(B \mid A)P(A)$ and this yields the well known *Bayes' rule*:

$$P(B \mid A) = \frac{P(A \mid B)P(B)}{P(A)}. \tag{2.3}$$

Bayes' rule conditioned on C reads

$$P(B \mid A, C) = \frac{P(A \mid B, C)P(B \mid C)}{P(A \mid C)}. \tag{2.4}$$

Formula (2.2) should be considered an axiom for probability calculus rather than a theorem. A justification for the formula can be found by counting frequencies: suppose we have m cats (C) of which n are brown (B), and i of the brown cats are Abyssinians (A). Then the frequency of As given B among the cats, $f(A \mid B, C)$, is $\frac{i}{n}$, the frequency of Bs, $f(B \mid C)$, is $\frac{n}{m}$, and the frequency of brown Abyssinian cats, $f(A, B \mid C)$ is $\frac{i}{m}$. Hence,

$$f(A \mid B, C)f(B \mid C) = f(A, B \mid C).$$

Likelihood

Sometimes $P(A \mid B)$ is called the *likelihood of B given A*, and it is denoted $L(B \mid A)$.

The reason for this is the following. Assume $B_1, \ldots, B_n$ are possible scenarios wilth an effect on the event A, and we know A. Then $P(A \mid B_i)$ is a measure of how likely it is that B_i is the cause. In particular, if all B_is have the same prior probability, Bayes' rule yields

$$P(B_i \mid A) = \frac{P(A \mid B_i)P(B_i)}{P(A)} = kP(A \mid B_i),$$

where k is independent of i.

2.3.3 Subjective probabilities

The justification in the previous section for the fundamental rule was based on frequencies. This does not mean that we only consider probabilities based on frequencies. Probabilities may also be completely subjective estimates of the certainty of an event.

A subjective probability may, for example, be my personal assessment of the chances of selling more than 2,000 copies of this book in 1997.

A way to assess this probability could be the following. I am given the choice between two gambles:

(1) if more than 2,000 copies are sold in 1997 I will receive \$100;

(2) I will by the end of 1997 be allowed to draw a ball from an urn with n red balls and $100 - n$ white balls. If my ball is red I will get \$100.

Now, if all balls in the urn are red I will prefer (2), and if all balls are white I will prefer (1). There is a number n for which the two gambles are equally attractive, and for this n, $\frac{n}{100}$ is my estimate of the probability of selling more than 2,000 copies of this book in 1997 (I shall not disclose the n to the reader).

For subjective probabilities defined through such ball drawing gambles the fundamental rule can also be proved.

2.3.4 Probability calculus for variables

As stated in Section 2.2, the nodes in a causal network are *variables* with a *finite number of mutually exclusive states*.

If A is a variable with states $a_1, \ldots, a_n$, then $P(A)$ is a probability distribution over these states:

$$P(A) = (x_1, \ldots, x_n) \qquad x_i \geq 0 \qquad \sum_{i=1}^{n} x_i = 1,$$

where x_i is the probability of A being in state a_i.

Notation. The probability of A being in state a_i is denoted $P(A = a_i)$ and denoted $P(a_i)$ if the variable is obvious from the context.

If the variable B has states $b_1, \ldots, b_m$, then $P(A \mid B)$ is an $n \times m$ table containing numbers $P(a_i \mid b_j)$ (see Table 2.1).

$P(A, B)$, the joint probability for the variables A and B, is also an $n \times m$ table. It consists of a probability for each configuration (a_i, b_j) (see Table 2.2).

When the fundamental rule (2.1) is used on variables A and B, then the procedure is to apply the rule to the $n \cdot m$ configurations (a_i, b_j):

$$P(a_i \mid b_j)P(b_j) = P(a_i, b_j).$$

This means that in the table $P(A \mid B)$, for each j the column for b_j is multiplied by $P(b_j)$ to obtain the table $P(A, B)$. If $P(B) = (0.4, 0.4, 0.2)$ then Table 2.2 is the result of using the fundamental rule on Table 2.1. When applied to variables, we use the same notation for the fundamental rule:

$$P(A \mid B)P(B) = P(A, B).$$

From a table $P(A, B)$ the probability distribution $P(A)$ can be calculated. Let a_i be a state of A. There are exactly m different events for which A is in state a_i, namely the mutually exclusive events $(a_i, b_1), \ldots, (a_i, b_m)$. Therefore, by axiom (ii)

$$P(a_i) = \sum_{j=1}^{m} P(a_i, b_j).$$

Table 2.1 An example of $P(A \mid B)$. Note that the columns sum to one.

	b_1	b_2	b_3
a_1	0.4	0.3	0.6
a_2	0.6	0.7	0.4

Table 2.2 An example of $P(A, B)$. Note that the sum of all entries is one.

	b_1	b_2	b_3
a_1	0.16	0.12	0.12
a_2	0.24	0.28	0.08

This calculation is called *marginalization* and we say that the variable B is marginalized out of $P(A, B)$ (resulting in $P(A)$). The notation is

$$P(A) = \sum_B P(A, B). \tag{2.5}$$

By marginalizing B out of Table 2.2 we get $P(A) = (0.4, 0.6)$.

The division in Bayes' rule (2.3) is treated in the same way as the multiplication in the fundamental rule (see Table 2.3).

2.3.5 Conditional independence

The blocking of transmission of evidence as described in Section 2.2.1 is, in the Bayesian calculus, reflected in the concept of *conditional independence.*The variables A and C are *independent given the variable* B if

$$P(A \mid B) = P(A \mid B, C). \tag{2.6}$$

This means that if the state of B is known then no knowledge of C will alter the probability of A.

Table 2.3 $P(B \mid A)$ as a result of applying Bayes' rule to Table 2.1 and $P(B) = (0.4, 0.4, 0.2)$.

	a_1	a_2
b_1	0.4	0.4
b_2	0.3	0.47
b_3	0.3	0.13

Remark. If condition B is empty, we simply say that A and C are independent.

Conditional independence appears in the cases of serial and diverging connections (see Figure 2.10).

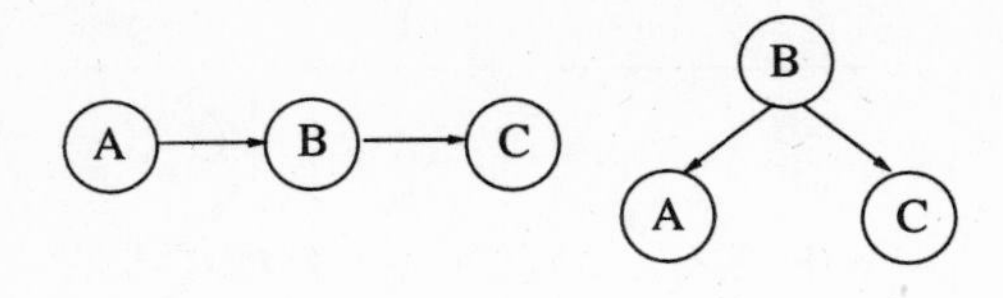

Figure 2.10 Examples where A and C are conditionally independent given B.

Definition (2.6) may look asymmetric; however, if (2.6) holds, then – by the conditioned Bayes' rule (2.4) – we get

$$P(C \mid B, A) = \frac{P(A \mid C, B)P(C \mid B)}{P(A \mid B)} = \frac{P(A \mid B)P(C \mid B)}{P(A \mid B)} = P(C \mid B).$$

The proof requires that $P(A \mid B) > 0$. That is, for states a, b with $P(A = a \mid B = b) = 0$ the calculation is not valid. However, for our considerations it does not matter; if B is in state b then the evidence $A = a$ is impossible and will not appear. So, why bother with the transmission of it?

2.3.6 Definition of Bayesian networks

Causal relations also have a quantitative side, namely their *strength*. This is expressed by attaching numbers to the links.

Let A be a parent of B. Using probability calculus it would be natural to let $P(B \mid A)$ be the strength of the link. However, if C is also a parent of B, then the two conditional probabilities $P(B \mid A)$ and $P(B \mid C)$ alone do not give any clue on how the impacts from A and B interact. They may co-operate or counteract in various ways. So, we need a specification of $P(B \mid A, C)$.

It may happen that the domain to be modelled contains feed-back cycles (see Fig. 2.11).

Feed-back cycles are difficult to model quantitatively (this is, for example, what differential equations are all about); for causal networks no calculus has been developed that can cope with feed-back cycles Therefore we require that the network does not contain cycles.

A Bayesian network consists of the following.

A set of *variables* and a set of *directed edges* between variables.

Each variable has a finite set of mutually exclusive states.

The variables together with the directed edges form a *directed acyclic graph* (DAG). (A directed graph is *acyclic* if there is no directed path $A_1 \to \cdots \to A_n$ such that $A_1 = A_n$.)

To each variable A with parents $B_1, \ldots, B_n$ there is attached a conditional probability table $P(A \mid B_1, \ldots, B_n)$.

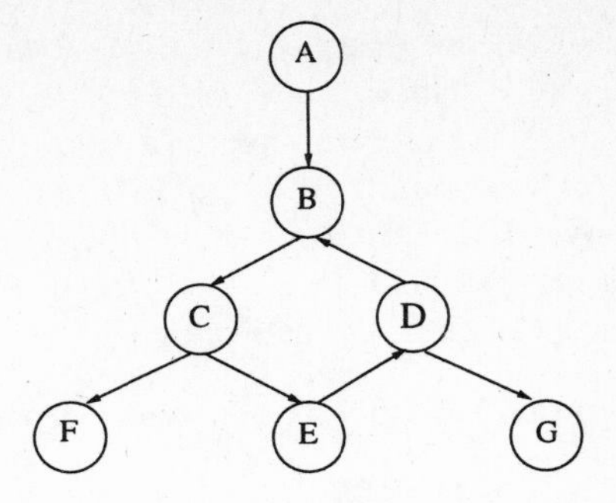

Figure 2.11 A directed graph with a feed-back cycle. This is not allowed in Bayesian networks.

Note that if A has no parents then the table reduces to unconditional probabilities $P(A)$. For the DAG in Figure 2.12 the prior probabilitiess $P(A)$ and $P(B)$ must be specified. It has been claimed that prior probabilities are an unwanted introduction of bias to the model, and calculi have been invented in order to avoid it. However, as discussed in Section 2.1.5, prior probabilities are necessary – not for mathematical reasons – but because prior certainty assessments are an integral part of human reasoning about certainty.

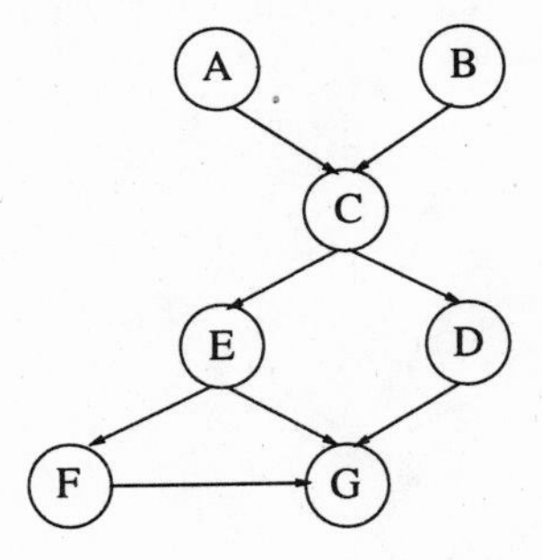

Figure 2.12 A directed acyclic graph (DAG). The probabilities to specify are $P(A)$, $P(B)$, $P(C \mid A, B)$, $P(E \mid C)$, $P(D \mid C)$, $P(F \mid E)$ and $P(G \mid D, E, F)$.

One of the advantages of Bayesian networks is that they *admit d-separation:* if A and B are d-separated in a Bayesian network with evidence e entered, then $P(A \mid B, e) = P(A \mid e)$. This means that you can use d-separation to read-off conditional independencies. We will use this fact without proof.

2.3.7 The chain rule

Let $U = (A_1, \ldots, A_n)$ be a universe of variables. If we have access to the joint probability table $P(U) = P(A_1, \ldots, A_n)$, then we can also calculate $P(A_i)$ as well as $P(A_i \mid e)$, where e is evidence (see Section 4.2). However, $P(U)$ grows exponentially with the number of variables, and U need not be very large before the table becomes intractably large. Therefore, we look for a more compact *representation* of $P(U)$: a way of storing information from which $P(U)$ can be calculated if needed.

A Bayesian network over U is such a representation. If the conditional independencies in the Bayesian network hold for U, then $P(U)$ can be calculated from the conditional probabilities specified in the network.

Theorem 2.1 (The chain rule.) *Let* BN *be a Bayesian network over*

$$U = \{A_1, \ldots, A_m\}.$$

Then the joint probability distribution $P(U)$ *is the product of all conditional probabilities specified in* BN*:*

$$P(U) = \prod_i P(A_i \mid pa(A_i))$$

where $pa(A_i)$ *is the parent set of* A_i.

Proof. (Induction in the number of variables in the universe U.)

If U consists of one variable then the theorem is trivial.

Assume the chain rule to be true for all networks consisting of $n-1$ variables, and let U be the universe for a DAG with n variables. Since the network is acyclic there is at least one variable A without children. Consider the DAG with A removed.

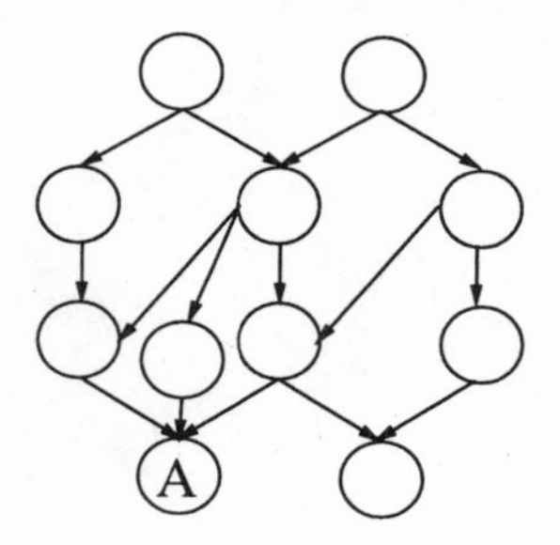

Figure 2.13 A DAG with n variables. If the variable A is removed, the induction hypothesis can be applied.

From the induction hypothesis we have that $P(U \setminus \{A\})$ is the product of all specified probabilities – except $P(A \mid pa(A))$.

By the fundamental rule we have

$$P(U) = P(A \mid U \setminus \{A\})P(U \setminus \{A\}).$$

Since A is independent of $U \setminus (\{A\} \cup pa(A))$ given $pa(A)$ (see Fig. 2.13), we get

$$P(U) = P(A \mid U \setminus \{A\})P(U \setminus \{A\}) = P(A \mid pa(A))P(U \setminus \{A\}).$$

The righthand side above is the product of all specified probabilities.

Table 2.4 Conditional probabilities for H and W.

	$I = y$	$I = n$
$H = y$	0.8	0.1
$H = n$	0.2	0.9

$P(H \mid I)$

	$I = y$	$I = n$
$W = y$	0.8	0.1
$W = n$	0.2	0.9

$P(W \mid I)$

Table 2.5 Joint probability table for $P(W, I)$ and $P(H, I)$.

	$I = y$	$I = n$
y	0.56	0.03
n	0.14	0.27

2.4 The examples revisited

In this section we apply the rules of probability calculus on the introductory examples. This is done to illustrate that probability calculus can be used to perform the reasoning in the examples – in particular explaining away. In Chapter 4 we give a general algorithm for probability updating in Bayesian networks. This algorithm makes the calculations considerably easier than those in this section.

2.4.1 Icy roads

(See Fig. 2.1.) For the quantitative modelling we need three probability assessments: $P(H \mid I)$, $P(W \mid I)$ and $P(I)$. The model in Figure 2.1 reflects that only knowledge of icy roads is relevant for H and W. We should then attach a certainty to I based on whatever knowledge may be available. In this case the police inspector has been looking out of the window and wondering whether the roads were icy. We let the probability for icy roads be 0.7.

Since both Holmes and Watson are bad drivers, we put the probability of a crash in the case of icy roads to 0.8, and the probability of a crash if the roads are not icy we put to 0.1 (they *are* bad drivers). An overview of the conditional probabilities is given in Table 2.4.

To calculate the initial probabilities for H and W we first use the fundamental rule (2.1) to calculate $P(W, I)$ and $P(H, I)$:

$$P(W = y, I = y) = P(W = y \mid I = y)P(I = y) = 0.8 \cdot 0.7 = 0.56.$$

Table 2.5 gives all four probabilities.

In order to get the probabilities for W and H we marginalize I out of Table 2.5 and get

$$P(W) = P(H) = (0.59, 0.41).$$

The information that Watson has crashed is now used to update the probability of

I. For this, Bayes' rule is used:

$$\begin{aligned} P(I \mid W = y) &= \frac{P(W = y \mid I)P(I)}{P(W = y)} \\ &= \frac{1}{0.59}(0.8 \cdot 0.7, 0.1 \cdot 0.3) \\ &= (0.95, 0.05). \end{aligned}$$

To update the probability of H, first we use the fundamental rule (2.1) to calculate $P(H, I)$ as shown in Table 2.6.

Table 2.6 Tables showing the calculation of $P(H, I)$.

	$I = y$	$I = n$			$I = y$	$I = n$
$H = y$	$0.8 \cdot 0.95$	$0.1 \cdot 0.05$	$=$	$H = y$	0.76	0.005
$H = n$	$0.2 \cdot 0.95$	$0.9 \cdot 0.05$		$H = n$	0.19	0.045

Finally, calculate $P(H)$ by marginalizing I out of $P(H, I)$. The result is

$$P(H) = (0.765, 0.235).$$

This is the quantitative effect of the information that Watson has crashed.

At last, when Inspector Smith is convinced that the roads are not icy, then $P(H \mid I = n) = (0.1, 0.9)$.

The calculation can be considered in a different way. First we calculate $P(H, I)$ and $P(W, I)$ (Table 2.5), and we have two joint probability tables with the variable I in common.

If evidence on W now arrives in the form of $P^*(W) = (0, 1)$, then

$$P^*(W, I) = P(I \mid W)P^*(W) = \frac{P(W, I)}{P(W)}P^*(W).$$

This means that the joint probability table for W and I is updated by multiplying by the new distribution and dividing by the old one. The multiplication consists of annihilating all entries with $W = n$. The division by $P(W)$ only has an effect on entries with $W = y$, so therefore the division is by $P(W = y)$.

Next, calculate $P^*(I)$ from $P^*(W, I)$ by marginalization, and use $P^*(I)$ to update $P(H, I)$

$$P^*(H, I) = \frac{P(H, I)}{P(I)} \cdot P^*(I)$$

and finally $P^*(H)$ is calculated by marginalizing $P^*(H, I)$.

2.4.2 Wet grass

(See Fig. 2.2.) Let the prior probabilities for R and S be $P(R) = (0.2, 0.8)$ and $P(S) = (0.1, 0.9)$. The remaining probabilities are listed in Table 2.7. First, calculate the prior probabilities for W and H by formulae (2.1) and (2.5). That is, first calculate $P(W, R)$ and then marginalize R out. The result is $P(W) = (0.36, 0.64)$.

Table 2.7 The probabilities for the *wet grass* example. The vectors (α, β) in the righthand table represent $(H = y, H = n)$.

	$R = y$	$R = n$
$W = y$	1	0.2
$W = n$	0	0.8

$P(W \mid R)$

	$R = y$	$R = n$
$S = y$	(1, 0)	(0.9, 0.1)
$S = n$	(1, 0)	(0, 1)

$P(H \mid R, S)$

The calculation of $P(H, R, S)$ follows the same scheme, only the product is

$$P(H, R, S) = P(H \mid R, S)P(R, S).$$

Since R and S are independent (see Fig. 2.2) we have (see Exercise 2.9)

$$P(H, R, S) = P(H \mid R, S)P(R)P(S).$$

The result is given in Table 2.8. Marginalizing R and S out of $P(H, R, S)$ yields $P(H) = (0.272, 0.728)$. We shall use the approach outlined at the end of Section 2.4.1. We have established joint probability tables for two of the clusters, (W, R) and (H, R, S), with the variable R in common.

Table 2.8 The prior probability table for $P(H, R, S)$. The vectors (α, β) in the table represent $(H = y, H = n)$.

	$R = y$	$R = n$
$S = y$	(0.02, 0)	(0.072, 0.008)
$S = n$	(0.18, 0)	(0, 0.72)

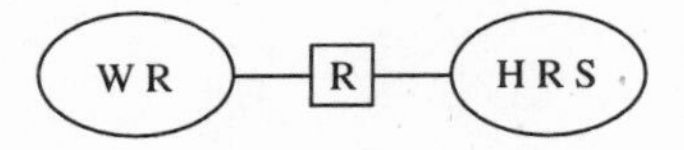

Figure 2.14 The clusters for the *wet grass* example. They communicate through the variable R.

The evidence $H = y$ is used to update $P(H, R, S)$ by annihilating all entries with $H = n$ and dividing by $P(H = y)$. Since the result shall be a probability table with all entries summing to one we need not calculate $P(H)$. After all entries with $H = n$ have been annihilated (Table 2.9), we simply normalize the table by dividing by the sum of the remaining entries (see Table 2.10).

The distributions $P^*(R)$ and $P^*(S)$ are calculated through marginalization of

$$P^*(H, R, S).$$

Table 2.9 $P(H, R, S)$ with all entries with $H = n$ annihilated.

	$R = y$	$R = n$
$S = y$	(0.02, 0)	(0.072, 0)
$S = n$	(0.18, 0)	(0, 0)

Table 2.10 The calculation of $P^*(H, R, S) = P(H, R, S \mid H = y)$.

	$R = y$	$R = n$			$R = y$	$R = n$
$S = y$	$\frac{1}{0.272}(0.02, 0)$	$\frac{1}{0.272}(0.072, 0)$	$\approx$	$S = y$	$(0.074, 0)$	$(0.264, 0)$
$S = n$	$\frac{1}{0.272}(0.18, 0)$	$\frac{1}{0.272}(0, 0)$		$S = n$	$(0.662, 0)$	$(0, 0)$

We get $P^*(R = y) = 0.736$ and $P^*(S = y) = 0.339$.

Use $P^*(R)$ to update $P(W, R)$ (see Table 2.11):

$$P^*(W, R) = P(W \mid R)P^*(R) = P(W, R)\frac{P^*(R)}{P(R)}.$$

Table 2.11 Calculation of $P^*(W, R) = P(W, R)\frac{P^*(R)}{P(R)}$.

	$R = y$	$R = n$			$R = y$	$R = n$
$W = y$	$0.2 \cdot \frac{0.736}{0.2}$	$0.16 \cdot \frac{0.264}{0.8}$	$=$	$W = y$	0.736	0.0528
$W = n$	0	$0.64 \cdot \frac{0.264}{0.8}$		$W = n$	0	0.2112

Now use $W = y$ to update the distribution for (W, R) (see Table 2.12). We get $P^{**}(R = y) = 0.93$.

We still have to calculate $P^{**}(S) = P(S \mid W = y, H = y)$. The result must reflect the explaining away effect; since the wet grass is explained by rain, the probability for $S = y$ should decrease to its initial value.

The calculation follows the same pattern. A message on $P^{**}(R)$ is sent from (W, R) to (H, R, S) (see Fig. 2.14),

$$P^{**}(H, R, S) = P^*(H, R, S)\frac{P^{**}(R)}{P^*(R)}.$$

By marginalizing we get $P^{**}(S = y) = 0.161$.

Table 2.12 $P^{**}(W, R) = P(W, R \mid W = y, H = y)$.

	$R = y$	$R = n$
$W = y$	$\frac{0.736}{0.7888}$	$\frac{0.0528}{0.7888}$
$W = n$	0	0

Table 2.13 $P^{**}(R, S) = P(R, S \mid H = y, W = y)$.

	$R = y$	$R = n$
$S = y$	0.094	0.067
$S = n$	0.839	0

The reason why the probability for sprinkler does not drop to the prior probability of 0.1 is that Dr Watson is a forgetful fellow who may have forgotten his sprinkler, and an explanation may be that both sprinklers have been forgotten. This is reflected in the probability $P(W = y \mid R = n) = 0.2$.

2.5 BOBLO

BOBLO is a system which helps in the verification of parentage for Jersey cattle through blood-type identification. The introduction of embryo transplantation technology and the increasing trade of semen and embryos have stressed the importance of proper pedigree registration, and therefore there is a need for sophisticated methods for individual identification and parentage control of cattle.

Heredity is determined by *genes* which are placed in chromosomes (see Fig. 2.15).

Figure 2.15 A pair of chromosomes. The pearls in the strings are loci.

Except for the sex chromosomes, chromosomes go in structurally identical pairs – one chromosome inherited from each parent. A chromosome may be considered as a string of genes. The places where the genes are positioned are called *loci*. Each gene has a particular locus of position and genes which can be placed at a particular locus are called *allels*. The pair of allels at a locus (one from each chromosome) is called a *genotype*, and the property determined by a genotype is called the *phenotype*.

For the blood group determination of cattle, ten different independent blood-group systems are used. These systems control 52 different blood-group *factors* which can be measured in a laboratory. In eight of these systems the blood-group determination is relatively simple (controlling from one to four blood-group factors only). However, the systems B- and C- are rather complicated, controlling respectively 26 and 10 of the above-mentioned 52 blood-group factors.

Heredity of blood type follows the normal genetic rules, however, the blood groups are attached to sets of loci rather than to single loci, and instead of allels the term *phenogroup* is used. So, for each blood group, a Bayesian network for inheritance will be as in Figure 2.16.

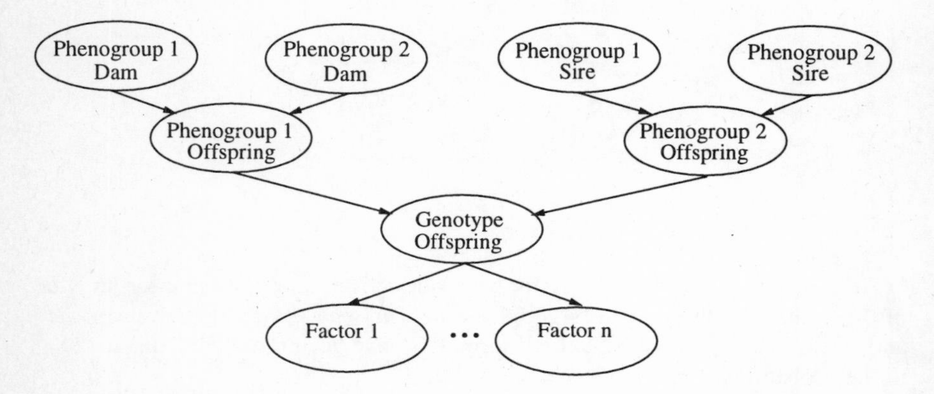

Figure 2.16 Heredity of blood type. From each parent one out of two phenotypes are chosen. This constitutes the genotype of the offspring, and the genotype determines a set of factors measurable in a laboratory (the phenotype).

If nothing is known of the phenogroups of the parents they are given a prior probability equal to the frequencies of the various phenogroups. Let us, for the example, suppose that there are three phenogroups f_1, f_2, f_3 with frequencies $(0.58, 0.1, 0.32)$ (this is the situation for the so-called *F-system*).

When a calf is registered, the parents are stated and their phenogroups are already registered. If the stated parents are the true parents we have no problems, but what if they are not so? Then we will say that the phenogroups of the true parents are distributed as the prior probabilities, that is $(0.58, 0.1, 0.32)$.

So, for modelling the part concerning possible parental errors, we can introduce a node *parental error* with states *both, sire, dam* and *no*, and with prior probabilities to be the frequency of parental errors. This leads to the Bayesian network in Figure 2.17.

The network model in BOBLO also has a part that models the risks of mistakes in the laboratory procedures (see Exercise 3.6). For now, assume that evidence on factors are entered directly to the nodes *factor*. It is assumed that the stated parents are so well known that their genotypes are known, and therefore the state of the variables *phenogroup stated d/s* is known.

Note how the impact of evidence flows from the *factor* nodes to the node *parental error*: it first flows to *phenogroup true d/s* (serial connections). Since evidence has

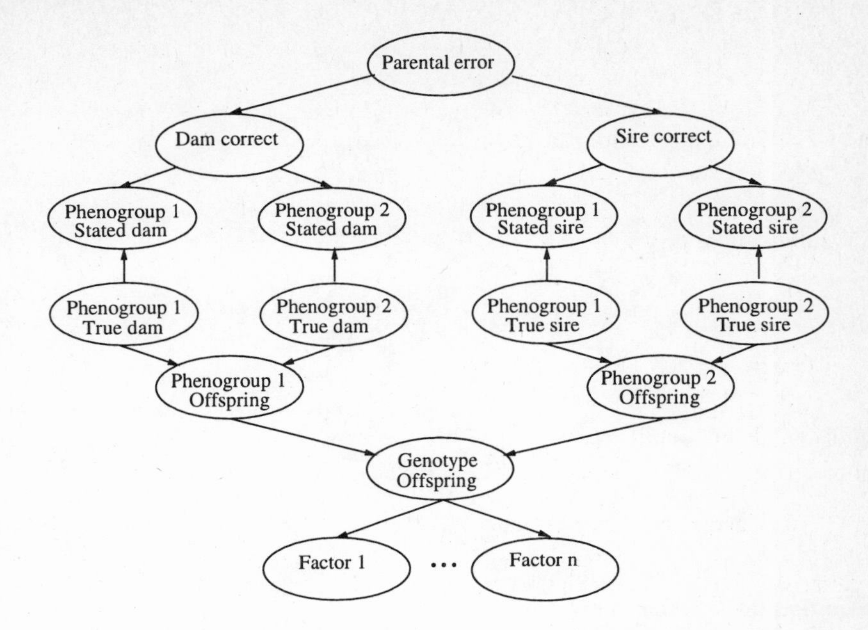

Figure 2.17 The part of BOBLO modelling parental error. Evidence is entered into the variables *factor* and *phenogroup stated d/s*. Evidence from *factor* is transmitted to *parental error* because *phenogroup stated* has received evidence.

been entered to *phenogroup stated d/s* the evidence is transmitted further to *dam correct* and *sire correct* (converging connections) to end in *parental error*.

BOBLO is an acronym for BOvine BLOod typing, and it has been in use at the Danish Blood Type Laboratory improving the accuracy of detecting parental errors (tests quantifying the improvement have not been finished).

2.6 Summary

d-separation in causal networks

Two variables A and B in a causal network are d-separated if for all paths between A and B there is an intermediate variable V such that either

- the connection is serial or diverging and the state of V is known or
- the connection is converging, and neither V nor any of Vs descendants have received evidence.

The fundamental rule for probability calculus

$$P(A \mid B, C)P(B \mid C) = P(A, B \mid C)$$

Bayes' rule

$$P(B \mid A, C) = \frac{P(A \mid B, C)P(B \mid C)}{P(A \mid C)}$$

Marginalization

$$P(A) = \sum_i P(A, b_i) = P(A, b_1) + \cdots + P(A, b_n)$$

Conditional independence

A and C are independent given B if $P(A \mid B) = P(A \mid B, C)$.

Definition of Bayesian networks

A Bayesian network consists of the following.

A set of *variables* and a set of *directed edges* between variables.

Each variable has a finite set of states.

The variables together with the directed edges form a *directed acyclic graph* (DAG).

To each variable A with parents $B_1, \ldots, B_n$ there is attached a conditional probability table $P(A \mid B_1, \ldots, B_n)$.

Admittance of d-separation in Bayesian networks

If A and B are d-separated in a Bayesian network with evidence e entered, then $P(A \mid B, e) = P(A \mid e)$.

The chain rule

Let *BN* be a Bayesian network over $U = \{A_1, \ldots, A_m\}$. Then the joint probability distribution $P(U)$ is the product of all conditional probabilities specified in *BN*:

$$P(U) = \prod_i P(A_i \mid pa(A_i)),$$

where $pa(A_i)$ is the parent set of A_i.

2.7 Bibliographical notes

The two Examples 2.1.2 and 2.1.4 are inspired by Pearl (1988). The concepts of causal network, d-connection, and the definition in Section 2.2.1 are due to Pearl (1986b) and Verma (1987). A proof that Bayesian networks admit d-separation can be found in Pearl (1988) or in Lauritzen (1996). Bayesian networks have a long history in statistics, and in the first half of the 1980s they were introduced to the field of expert systems through work by Pearl (1982) and Spiegelhalter & knill-Jones (1984). BOBLO is documented in Rasmussen (1995a,b).

Exercises

Exercise 2.1 Show that d-connectedness is *symmetric* (if A is d-connected to B, then B is d-connected to A).

Give an example proving that d-connectedness is not *transitive* (A d-connected to B and B d-connected to C, but A and C are not d-connected).

Exercise 2.2 In the graphs below determine which variables are d-connected to A.

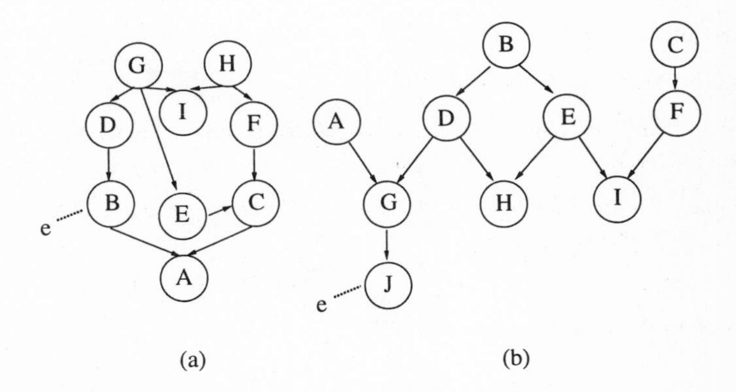

Figure for Exercise 2.2

Exercise 2.3 Let A be a variable in a DAG. Assume that the following variables are instantiated: the parents of A, the children of A, the spouses of A (variables that share a child with A).

Show that A is d-separated from the remaining uninstantiated variables.

Exercise 2.4 Let D_1 and D_2 be DAGs over the same variables. D_1 is an *I-submap* of D_2 if all d-separation properties of D_1 also hold for D_2. If, also, D_2 is an I-submap of D_2, they are sid to be *I-equivalent*.

Which of the four DAGs in the figure below are I-equivalent?

Table 2.14 Table for Exercise 2.5.

	b_1	b_2	b_3
a_1	0.05	0.10	0.05
a_2	0.15	0.00	0.25
a_3	0.10	0.20	0.10

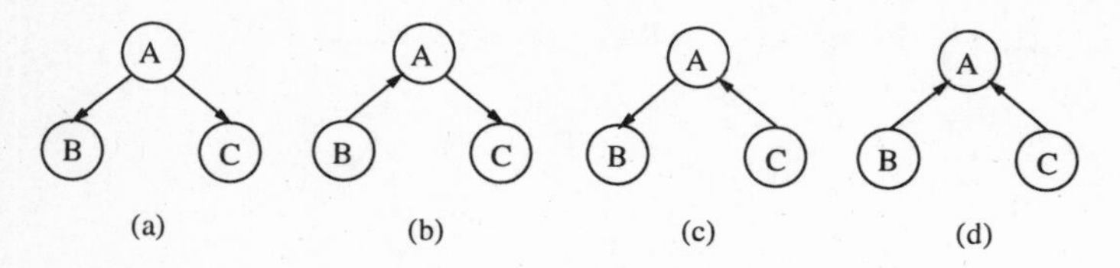

Figure for Exercise 2.4.

Exercise 2.5 Calculate $P(A)$, $P(B)$, $P(A \mid B)$, and $P(B \mid A)$ from Table 2.14.

Table 2.15 $P(A, B, C)$ for Exercise 2.6.

	b_1	b_2
a_1	(0.006, 0.054)	(0.048, 0.432)
a_2	(0.014, 0.126)	(0.032, 0.288)

Table 2.16 Conditional probability tables for Exercise 2.7.

	a_1	a_2		a_1	a_2
b_1	0.2	0.3	c_1	0.5	0.6
b_2	0.8	0.7	c_2	0.5	0.4
	$P(B \mid A)$			$P(C \mid A)$	

Exercise 2.6 In Table 2.15, a joint probability table for the binary variables A, B, and C is given.

(i) Calculate $P(B, C)$ and $P(B)$.

(ii) Are A and C independent given B?

Exercise 2.7 The DAG (a) in Exercise 2.4 has $P(A) = (0.1, 0.9)$ and the conditional probability given in Table 2.16.

Calculate $P(A, B, C)$.

Exercise 2.8 Perform a Bayesian calculation of the reasoning in Section 2.1.4 (earthquake or burglary). Use the probabilities in Table 2.17 and $P(B) = (0.01, 0.99)$, $P(E) = (0.001, 0.999)$.

Table 2.17 Tables for Exercise 2.8. Probabilities for radio and alarm.

	$E = y$	$E = n$
$R = y$	0.95	0.01
$R = n$	0.05	0.99

$P(R \mid E)$

	$B = y$	$B = n$
$E = y$	(0.98, 0.02)	(0.95, 0.05)
$E = n$	(0.95, 0.05)	(0.03, 0.97)

$P(A \mid B, E)$

Exercise 2.9 Let $P(c_i \mid b_j) \neq 0$ for all i, j. Prove that A and C are independent given B if and only if $P(A, C \mid B) = P(A \mid B)P(C \mid B)$.

Chapter 3

Building models

Bayesian networks create a very efficient language for building models of domains with inherent uncertainty. However, as can be seen from the calculations in Section 2.4, it is a tedious job to perform evidence transmission even for very simple Bayesian networks. Fortunately, software tools which can do the calculation job for us are available. Several commercial products exist containing both an editor for Bayesian networks and a runtime module which takes care of evidence transmission. In the rest of this book we assume that the reader has access to the HUGIN system provided by the diskette attached to the book, or to any other Bayesian network programming environment.

Therefore we can start by concentrating on how to use Bayesian networks in model building and defer a presentation of the methods for probability updating to Chapter 4.

In Section 3.1 we examine, through three examples, the considerations when determining the structure of a Bayesian network model. Section 3.2 gives examples of estimation of the conditional probabilities. The examples cover theoretically well-founded probabilities as well as probabilities taken from data bases and purely subjective estimates. Section 3.3 gives several modelling tricks to use when the amount of numbers to acquire is overwhelming. In Section 3.4 we touch upon methods for learning structure from a data base and for adapting the conditional probabilities to incoming cases.

Finally we describe the system *Child*.

3.1 Catching the structure

3.1.1 Family out?

> When I go home at night, I want to know if my family is home before I try the doors. (Perhaps the most convenient door to enter is double locked when nobody is home.) Now, often when my wife leaves the house she turns on an outdoor light. However, she sometimes turns on this light if she is expecting a guest. Also, we have a dog. When

nobody is home, the dog is put in the back yard. The same is true if the dog has bowel trouble. Finally, if the dog is in the back yard, I will probably hear her barking, but sometimes I can be confused by other dogs barking.

The first thing to have in mind when organizing a Bayesian model for a decision support system is that its purpose is to give estimates of certainties for events which are *not observable* (or only observable at an unacceptable cost). So, the primary task in model building is to identify these events. We call them *hypothesis events*.

Here we have two hypothesis events, namely *family at home* and *family out*.

Now, the hypothesis events have to be organized into a set of variables. A variable incorporates an exhaustive set of mutually exclusive events. That is, for each variable precisely one of its events is true.

Here it is very easy to organize the events into one variable *F-out?* with states y and n.

The next thing to have in mind is that in order to come up with a certainty estimate, we should provide some *information channels*. So, the task is to identify the types of achievable information which may reveal something about the state of some hypothesis variable. This is also done by establishing certain variables, *information variables*, such that a piece of information corresponds to a statement about the state of an information variable. Typically, the information will be a statement that a particular information variable is in a particular state; but also more soft statements are allowed.

Here, the information variables are *L-on?* (light on) with states y and n and *H-bark?* (hear bark) also with states y and n.

Now it is time to consider the causal structure between the variables. At this stage we need not worry about how information is transmitted through the network. The only thing to worry about is which events have a direct causal impact on other events.

In this example it is clear that *F-out?* has an impact on *L-on?* as well as on *H-bark?*, and that there is no causal relation between *H-bark?* and *L-on?*.

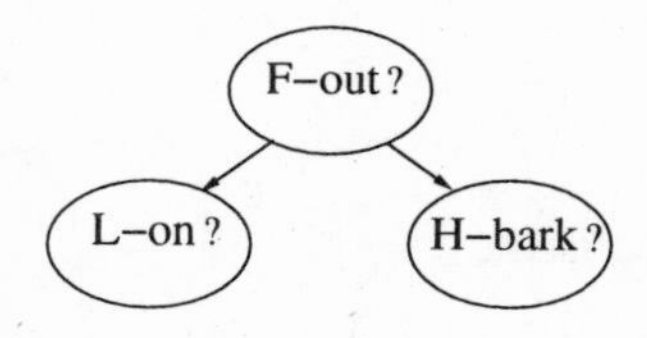

Figure 3.1 A causal structure for *family-out?*

We may stop with the model in Figure 3.1 and start specifying the probabilities $P(F\text{-}out)$, $P(H\text{-}bark? \mid F\text{-}out?)$ and $P(L\text{-}on? \mid F\text{-}out?)$. We will defer the remaining treatment of this example to the section on specification of the probabilities (Section 3.2.4).

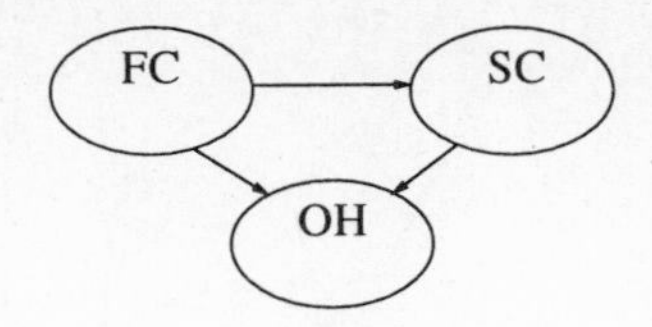

Figure 3.2 An oversimplified structure for the poker game. The variables are *FC* (first change), *SC* (second change), and *OH* (opponent's hand).

3.1.2 A simplified poker game

In this poker game each player receives three cards and is allowed two rounds of changing cards. In the first round you may discard any number of cards from your hand and get replacements from the pack of cards. In the second round you may discard at most two cards. After the two rounds of card changing, I am interested in an estimate of my opponent's hand.

The hypothesis events are the various types of hands in the game. They may be classified in the following way (in increasing rank): nothing special, 1 ace, 2 of the same value, 2 aces, flush (3 of a suit), straight (3 of consecutive value), 3 of the same value, straight flush. Ambiguities are resolved according to rank. This is of course a simplification, but you often have to do so when modelling. The hypothesis events are collected into one hypothesis variable *OH* (opponent's hand) with the classes given above as states.

The only information to acquire is the number of cards the player discards in the two rounds. (By saying so, we again are making an approximation. The information on the cards you have seen is relevant for your opponent's hand. If, for example, you have seen three aces then he cannot have two aces.)

So, the information variables are *FC* (first change) with states *0, 1, 2, 3* and *SC* (second change) with states *0, 1, 2*.

A causal structure for the information variables and the hypothesis variable could be as in Figure 3.2.

However, this structure will leave us with no clue as to how to specify the probabilities.

What we need are variables describing the opponent's hands in the process: the initial hand *OH0* and the hand *OH1* after the first change of cards. The causal structure will then be as in Figure 3.3.

To determine the states of *OH0* and *OH1* we have to produce a classification which is relevant for the determination of the states of the children (*FC* and *OH1*, say). We may let *OH0* and *OH1* have the following states: *nothing special, 1 ace, 2 of consecutive value, 2 of a suit, 2 of the same value, 2 of a suit and 2 of consecutive value, 2 of a suit and 2 of the same value, 2 of consecutive value and 2 of the same value, flush, straight, 3 of the same value, straight flush.*

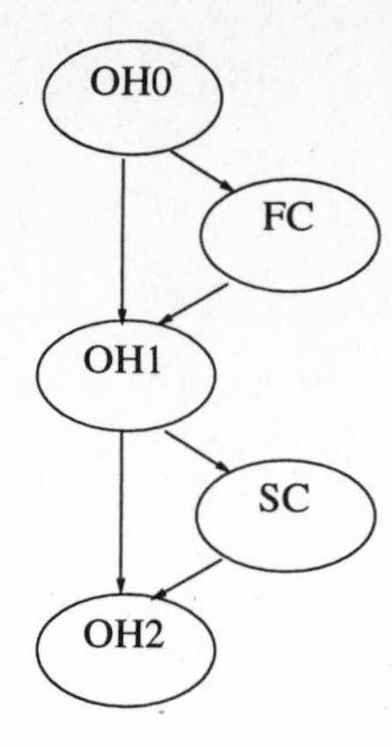

Figure 3.3 A structure for the poker game. The two mediating variables *OH0* and *OH1* are introduced. *OH2* is the variable for my opponent's final hand.

We defer further discussion of the classification to the section on specifying the probabilities (Section 3.2.2).

Variables in a model which are neither hypothesis variables nor information variables are called *mediating variables*. The decision on how to incorporate mediating variables is mainly a question of convenience. Usually mediating variables will ease the acquisition of conditional probabilities and thereby also increase the precision of the model. On the other hand there is a risk of increasing the complexity to a level which may jeopardize performance.

Another point is that it may happen that two variables – A and B – are dependent, but this dependence does not factor through any of the other variables. On the other hand, there is no obvious causal direction on the dependence. This should be taken as an indication that a mediating variable should be introduced as a parent of A and B. The next example illustrates this point.

3.1.3 Insemination

> Six weeks after insemination of a cow there are three tests for the result: blood test (BT), urine test (UT) and scanning (Sc). The results of the blood test and the urine test are mediated through the hormonal state (Ho) which is affected by a possible pregnancy (Pr). (This is a constructed example.)

A model will be like the one shown in Figure 3.4.

For both the blood test and the urine test there is a risk that a pregnancy does not show after six weeks. This is due to the fact that the change in the hormonal state may be too weak. Therefore, given pregnancy, the variables *BT* and *UT* are dependent.

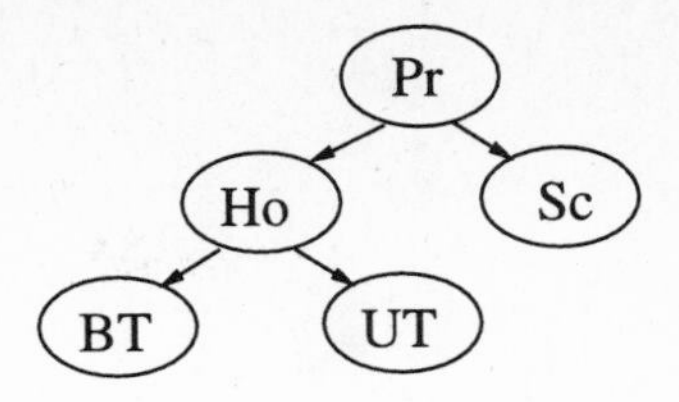

Figure 3.4 A model for test of pregnancy (Pr). Both the blood test (BT) and the urine test (UT) measure the hormonal state (Ho).

If we did not include the mediating variable, the model would be the one shown in Figure 3.5.

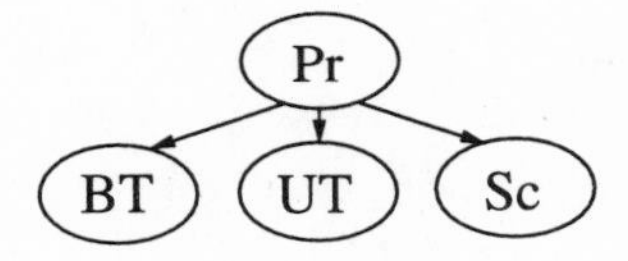

Figure 3.5 The pregnancy model without the *hormonal state* variable.

This model assumes the two tests to be independent given Pr.

If the model in Figure 3.5 is used for diagnosing a possible pregnancy, a negative outcome of both the blood test and the urine test will be counted as two independent pieces of evidence and therefore overestimate the probability for the insemination to have failed. (See Exercise 3.1.)

3.1.4 Simple Bayes models

The first Bayesian diagnostic systems were constructed through the following procedure.

- Let the possible diseases be collected into one hypothesis variable H with prior probability $P(H)$.
- For all information variables I, acquire the conditional probability $P(I \mid H)$ (the likelihood of H given I).
- For any set of findings $f_1, \ldots, f_n$ on the variables $I_1, \ldots, I_n$ calculate the product $L(H \mid f_1, \ldots, f_n) = P(f_1 \mid H)P(f_2 \mid H) \cdot \cdots \cdot P(f_n \mid H)$. This product is called the *likelihood* for H given $f_1, \ldots, f_n$. The posterior probability for H is calculated as $\mu P(H)L(H \mid f_1, \ldots, f_n)$, where μ is a normalization constant.

The calculations above reflect the simple model shown in Figure 3.6. (See Exercise 3.2.)

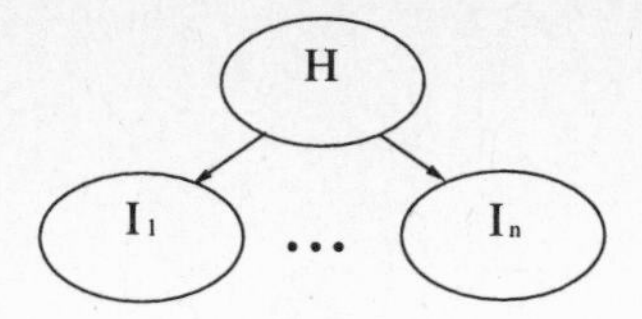

Figure 3.6 A simple Bayes model.

The model assumes that the information variables are independent given the hypothesis variable. As can be seen from the insemination example, the assumption need not hold, and if the model is used anyway, the conclusions may be misleading.

3.1.5 Causality

In the examples presented in the previous section there was no problem in establishing the links and their direction. However, you cannot expect this part of the modelling to always go smoothly.

First of all, causal relations are not always obvious – recall the debate on whether or not smoking causes lung cancer, or whether a person's sex has an impact on their abilities in the technical sciences. Furthermore, causality is not a well understood concept: is a causal relation a property of the real world, or, rather, is it a concept in our minds helping us to organize our perception of the world? We shall, however, not go into the scientific debate on causality and how to discover causal relations.

One point only. Causality has to do with actions where the state of the world is changed: you may, for example, find yourself confronted with two correlated variables A and B, but you cannot determine a direction. If you observe the state of A you will change your belief of B, and vice versa. A good test is then to imagine that some outside agent *fixes* the state of A. If this does not make you change the belief of B, then A is not a cause of B.

On the other hand, if this imagined test indicates a causal arrow in both directions, then you should look for an event which has a causal impact on both A and B. If C is such a candidate, then check whether A and B become independent given C.

3.2 Determining the conditional probabilities

The basis for the conditional probabilities in a Bayesian network can have different epistemological status ranging from well-founded theory over frequencies in a data base to subjective estimates. We shall give examples of each type.

3.2.1 Stud farm

> The stallion Brian has sired Dorothy with the mare Ann and sired Eric with the mare Cecily. Dorothy and Fred are the parents of Henry, and Eric has sired Irene with Gwenn. Ann is the mother of both Fred and Gwenn, but their fathers are in no way related. The colt John with the

Table 3.1 P(child | father, mother) for genetic inheritance. The numbers (α, β, γ) are the child's probabilities for *(aa, aA, AA)*.

	aa	aA	AA
aa	(1, 0, 0)	(0.5, 0.5, 0)	(0, 1, 0)
aA	(0.5, 0.5, 0)	(0.25, 0.5, 0.25)	(0, 0.5, 0.5)
AA	(0, 1, 0)	(0, 0.5, 0.5)	(0, 0, 1)

parents Henry and Irene has been born recently; unfortunately, it turns out that John suffers from a life threatening hereditary disease carried by a recessive gene. The disease is so serious that John is displaced instantly, and as the stud farm wants the gene out of production, Henry and Irene are taken out of breeding. What are the probabilities for the remaining horses to be carriers of the unwanted gene?

The geneological structure for the horses is given in Figure 3.7.

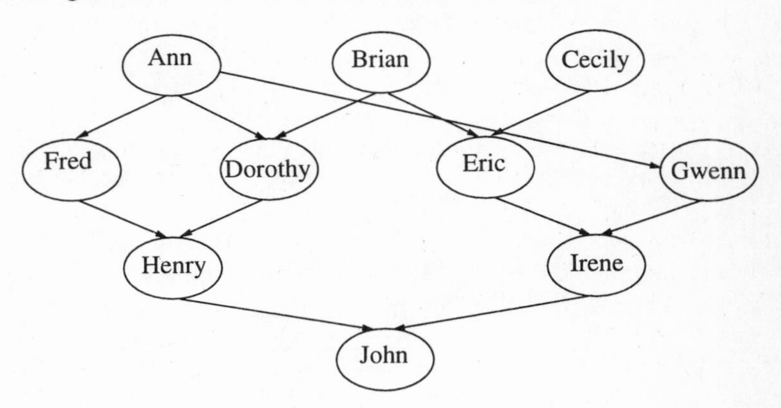

Figure 3.7 Geneological structure for the horses in the stud farm.

The only information variable is John. Before the information on John is acquired he may have three genotypes: he may be sick (aa), a carrier (aA), or he may be pure (AA). The hypothesis events are the genotypes of all other horses in the stud farm.

The conditional probabilities for inheritance are both empirically and theoretically well studied, and the probabilities are as shown in Table 3.1. The inheritance tables could be as Table 3.1. However, for all horses except John we have additional knowledge. Since they are in production they cannot be of type aa. A way to incorporate this would be to build a Bayesian network where all inheritance is modelled in the same way and afterwards enter the findings that all horses but John are not aa. It is also possible to calculate the conditional probabilities directly. If

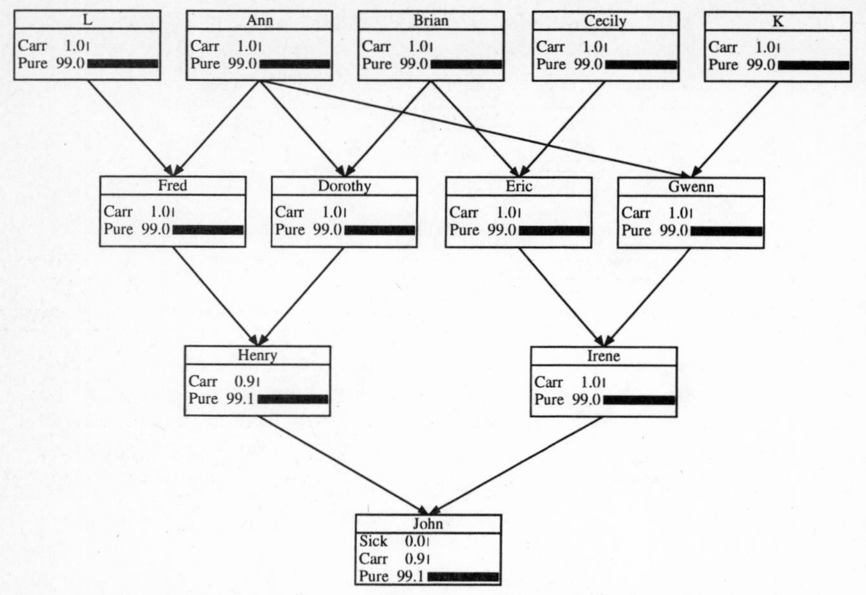

Figure 3.8 The *stud farm* model with initial probabilities. (HUGIN dump.)

we first consider inheritance from parents which may only be of genotype aA or AA, we get Table 3.2.

Table 3.2 $P(\text{child} \mid \text{father, mother})$ when the parents are not sick.

	aA	AA
aA	(0.25, 0.5, 0.25)	(0, 0.5, 0.5)
AA	(0, 0.5, 0.5)	(0, 0, 1)

The table for John is the same as in Table 3.2. For the other horses we know that aa is impossible. This is taken care of by removing the state aa from the distribution and normalizing the remaining distribution. For example $P(\text{child} \mid aA, aA) = (0.25, 0.5, 0.25)$, but since aa is impossible we get the distribution $(0, 0.5, 0.25)$ which is normalized to $(0, 0.67, 0.33)$. The final result is shown in Table 3.3.

In order to deal with Fred and Gwenn we introduce the two unknown fathers, I and K, as mediating variables and assume that they are not sick. For the horses at the top of the network we shall specify prior probabilities. This will be an estimate of the frequency of the unwanted gene, and there is no theoretical way to come up with it. Let us assume that the frequency is so that the prior belief of a horse being a carrier is 0.01.

Table 3.3 $P(\text{child} \mid \text{father, mother})$ with aa removed.

	aA	AA
aA	(0.67, 0.33)	(0.5, 0.5)
AA	(0.5, 0.5)	(0, 1)

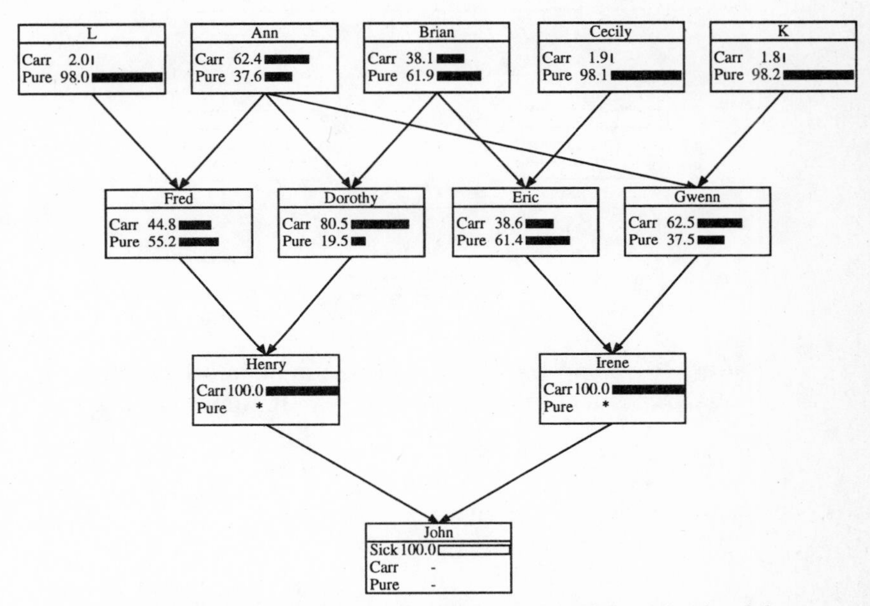

Figure 3.9 *Stud farm* probabilities given that *John* is sick. (HUGIN dump.)

In Figure 3.8 the final model with initial probabilities is shown, Figure 3.9 gives the posterior probabilities given John is aa, and in Figure 3.10 you can see the posterior probabilities with the prior beliefs at the top changed to 0.0001. Note that the sensitivity to the prior beliefs is very small for the horses where the posterior probability for *carrier* is well beyond zero, e.g. Ann and Brian.

3.2.2 Conditional probabilities for the poker game

In the *stud farm* example the conditional probabilities were mainly established through theoretical considerations. This should also be attempted for the model of the poker game developed in Section 3.1.2, but it cannot be carried through entirely.

Consider, for example, $P(FC \mid OHO)$. It is not possible to give probabilities which are valid for any opponent. It is heavily dependent on the opponent's insight, psychology and game strategies. We shall assume the following strategy.

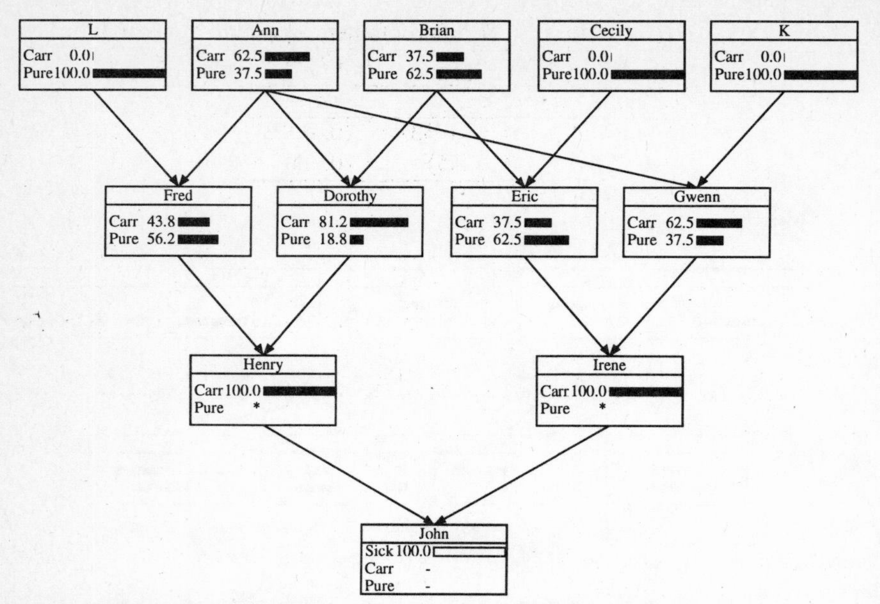

Figure 3.10 *Stud farm* probabilities with prior probabilities for top variables changed to (0.0001, 0.9999). (HUGIN dump.)

If nothing special (*no*), then change 3.

If 1 ace (*1 a*), then keep the ace.

If 2 of consecutive value (*2 cons*) or 2 of a suit (*2 s*) or 2 of the same value (*2 v*) then discard the third card.

If 2 of a suit and 2 of consecutive value, then keep 2 of a suit. (This strategy could be substituted by a random strategy for either keeping 2 of a suit or 2 of consecutive value.)

If 2 of a suit and 2 of the same value or 2 of consecutive value and 2 of the same value, then keep the 2 of the same value,.

If flush (*fl*), straight (*st*), 3 of the same value (*3 v*) or straight flush (*sfl*), then keep it.

Based on the strategy above, a logical link between FC and $OH0$ is established. Note that the strategy makes the states for combined hands redundant. They play no role, and therefore we remove them.

The strategy for $P(SC \mid OH1)$ is the same except that in the case of *no*, only 2 cards are discarded.

The remaining probabilities to specify are $P(OH0), P(OH1 \mid OH0, FC)$ and $P(OH2 \mid OH1, SC)$.

Table 3.4 $P(OH1 \mid OH0, FC)$ for the non-obvious parent configurations.

		(OH0, FC)				
		(*no, 3*)	(*1 a, 2*)	(*2 cons, 1*)	(*2 s, 1*)	(*2 v, 1*)
	no	0.1583	0	0	0	0
	1 a	0.0534	0.1814	0	0	0
	2 cons	0.0635	0.0681	0.3470	0	0
	2 s	0.4659	0.4796	0.3674	0.6224	0
OH1	*2 v*	0.1694	0.1738	0.1224	0.1224	0.9592
	fl	0.0494	0.0536	0	0.2143	0
	st	0.0353	0.0383	0.1632	0.0307	0
	3 v	0.0024	0.0026	0	0	0.0408
	sfl	0.0024	0.0026	0	0.0102	0

P(*OH0*). The states are (*no, 1 a, 2 cons, 2 s, 2 v, fl, st, 3 v, sfl*).

Through various (approximated) combinatorial calculations the prior probability distribution is found to be

$P(OH0) = (0.1672, 0.0445, 0.0635, 0.4659, 0.1694, 0.0494, 0.0353, 0.0024, 0.0024)$

P(*OH1* | *OH0*, *FC*). Due to the logical links between *OH0* and *FC* it is sufficient to consider only nine out of the possible 36 parent configurations, namely (*no, 3*), (*1 a, 2*), (*2 cons, 1*), (*2 s, 1*), (*2 v, 1*), (*fl, 0*), (*st, 0*), (*3 v, 0*), (*sfl, 0*). The last four are obvious. In Table 3.4 the results of approximate combinatorial calculations are given.

The probabilities for the remaining parent configurations may be whatever convenient. So, put, for example, $P(OH1 \mid 3\ v, 1) = (1, 0, 0, 0, 0, 0, 0, 0, 0)$.

P(*OH2* | *OH1*, *SC*). First a table $P(OH2' \mid OH1, SC)$ similar (but not identical in the numbers) to Table 3.4 can be calculated. However, the states of OH2′ are not the ones we are interested in. We are interested in the *value* of the hand and a state like *2 cons* is of no value unless one of them is an ace. Therefore, the probabilities for the states of *OH2′* are transformed to probabilities for *OH2*. For the transformation, the following rules are used:

$$1\ a = 1\ a + \frac{1}{6}(2\ cons + 2\ s)$$

$$no = no + \frac{5}{6}(2\ cons + 2\ s).$$

The probabilities of *2 a* are calculated specifically. The resulting probabilities are given in Table 3.5.

Using a model like the one in Figure 3.3 and with the conditional probability tables specified in this section, we have established a model for assisting a (novice) poker player. However, if my opponent knows that I use the system he may choose

Table 3.5 $P(OH2 \mid OH1, SC)$ for the non-obvious configurations.

		(OH1, Sc)				
		(no, 2)	(1 a, 2)	(2 cons, 1)	(2 s, 1)	(2 v, 1)
	no	0.5613	0	0.5903	0.5121	0
	1 a	0.1570	0.7183	0.1181	0.1024	0
	2 v	0.1757	0.0667	0.1154	0.1154	0.8838
OH2	2 a	0.0055	0.1145	0.0096	0.0096	0.0736
	fl	0.0559	0.0559	0	0.2188	0
	st	0.0392	0.0392	0.1666	0.0313	0
	3 v	0.0027	0.0027	0	0	0.0426
	sfl	0.0027	0.0027	0	0.0104	0

to change his strategies. His goal is to win rather than to obtain good hands, and he therefore may choose a strategy that makes me overestimate his hand. For instance, it seems a good strategy to discard two cards instead of three in the case of *no*. I will be convinced that he has an ace, and his chances for a good hand are not substantially reduced. We will return to this point in Chapter 6 on decision making.

3.2.3 Transmission of symbol strings

A language L over 2 symbols (**a**, **b**) is transmitted through a channel. Each word is surrounded by the delimiter symbol c. In the transmission some characters may be corrupted by noise and be confused with others.

A five-letter word is transmitted. Give a model which can determine the probabilities for the transmitted symbols given the received symbols.

There are five hypothesis variables $T_1, \ldots, T_5$ with states a and b and five information variables $R_1, \ldots, R_5$ with states a, b, c. Besides, mediating variables for the delimiters before and after the word may be considered. There is a causal relation from T_i to R_i. Furthermore, there may also be a relation from T_i to $T_{i+1}(i = 1, \ldots, 4)$. You could also consider more involved relations from pairs of symbols to symbols, but for now we refrain from that. The structure is given in Figure 3.11.

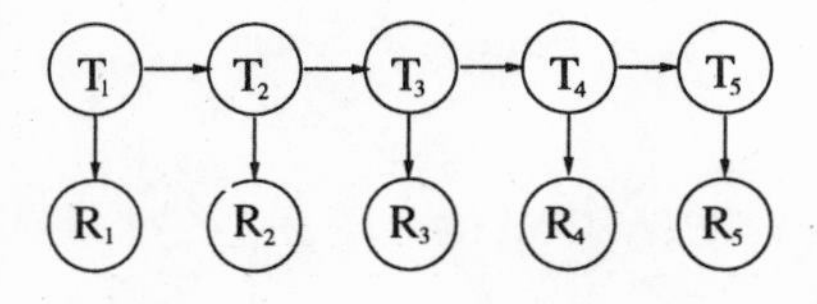

Figure 3.11 A model for symbol transmission. T_i are the symbols transmitted, R_i are the symbols received.

The conditional probabilities can be established through experience. The proba-

Table 3.6 $P(R \mid T)$ under transmission.

	$T = a$	$T = b$
$R = a$	0.80	0.15
$R = b$	0.10	0.80
$R = c$	0.10	0.05

Table 3.7 Frequencies of five-letter words in L. The word **abaab** for example has frequency 0.040.

	Last 3 letters							
First 2 letters	**aaa**	**aab**	**aba**	**abb**	**baa**	**bab**	**bba**	**bbb**
aa	0.017	0.021	0.019	0.019	0.045	0.068	0.045	0.068
ab	0.033	0.040	0.037	0.038	0.011	0.016	0.010	0.015
ba	0.011	0.014	0.010	0.010	0.031	0.046	0.031	0.045
bb	0.050	0.060	0.056	0.057	0.016	0.023	0.015	0.023

bilities $P(R_i \mid T_i)$ will be based on statistics describing the frequencies of confusion. Let Table 3.6 be the result.

You may obtain the probabilities $P(T_{i+1} \mid T_i)$ by investigating the five-letter words in L. What is the frequency of the first letter? What is the frequency of the second letter given that the first letter is **a**, etc. You can refine this frequency analysis by also taking the frequency of the words into consideration. Let Table 3.7 be the result of a frequency analysis.

You can calculate the required probabilities from Table 3.7. The prior probabilities for T_1 are (0.5, 0.5). Table 3.8 gives two conditional probabilities.

An alternative model would be to have a hypothesis variable, *Word*, with 32 states and with Table 3.7 as prior probabilities (see Fig. 3.12).

This is manageable because of the small amount of five-letter words over {**a**, **b**}; but if the alphabet had 24 symbols and six-letter words were considered the number of states in *Word* would become intractably large. On the other hand, the model of Figure 3.11 may be too simple to catch the dependencies in Table 3.7. So, the task really is to analyze the table in order to find the simplest structure describing

Table 3.8 Two conditional probabilities for five-letter words in L.

	a	b
a	0.6	0.4
b	0.4	0.6
	$P(T_2 \mid T_1)$	

	a	b
a	0.4	0.74
b	0.6	0.26
	$P(T_3 \mid T_2)$	

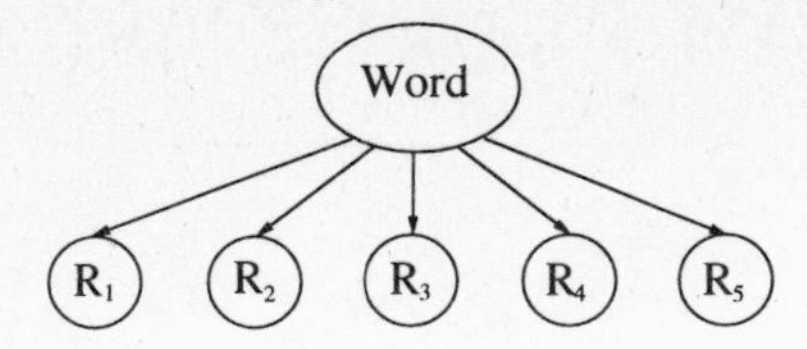

Figure 3.12 An alternative model for symbol transmission. *Word* is the possible transmitted words.

it. There are methods for doing this, and we shall revert to these in Section 3.4.

3.2.4 Family out?

The estimation of the conditional probabilities for the example introduced in Section 3.1.1 is a rather subjective task. Therefore, the model in Figure 3.1 may be changed in order to get better founded probabilities.

P(*F-out?*). I should give an estimate of how often my family is out when I return from work. Out of the five working days a week it only happens once, so I put $P(F\text{-}out?) = (0.2, 0.8)$.

P(*L-on?* | *F-out?*). As a rule the light is on when the family is out. Though I do not recall, they may have forgotten it now and then. Consequently, I put $P(L\text{-}on? \mid F\text{-}out? = y) = (0.99, 0.01)$.

As a rule, the light is off when the family is at home. However, when we expect guests the light is on, and this happens three times a month. We can either include this in the probability or we can add a mediating variable *Exp-g?* to make it explicit. Which one to use is partly a matter of taste and partly a question of how easy the required probabilities would be to estimate. Here, we hide the exception to the rule in the probabilities and put $P(L\text{-}on? \mid F\text{-}out? = n) = (0.1, 0.9)$.

There are several reasons why I can hear barking in the case of my family being at home as well as being out. In order to sort this out we introduce the mediating variable *D-out?* (see Fig. 3.13).

Again, to estimate $P(D\text{-}out? \mid F\text{-}out?)$ several factors are involved. Sometimes the dog is out due to *F-out?*, sometimes due to bowel problems and sometimes just because she wants to be. We introduce a mediating variable *BP?* (bowel problems) with prior probability (0.05, 0.95). Basically I would say that the dog is outside 20% of the time "just because"; so $P(D\text{-}out? \mid F\text{-}out? = n, BP? = n) = (0.2, 0.8)$. I also estimate that in 15% of the cases where my family is out they forget to let the dog out, and in 95% of the cases where the dog has bowel problems they let her out. What if the family is out and the dog has bowel problems? We can say that there is a "background" probability of 0.2 that the dog is out. Out of the remaining 80%, 85% of the time the dog is out due to the family being out, and in the remaining 12% , 95% of the time the dog is out due to bowel problems

Table 3.9 $P(D\text{-}out? = y \mid F\text{-}out?, BP?)$.

	$BP? = y$	$BP? = n$
$F\text{-}out? = y$	0.994	0.880
$F\text{-}out? = n$	0.960	0.200

(see Table 3.9). We shall revert to this line of reasoning in Section 3.3.2 on "noisy or". To estimate $P(H\text{-}bark? \mid D\text{-}out?)$ we may also perform a detailed analysis introducing my neighbour's annoying dog who is always out howling incessantly; but we may also make him implicit in the table such that $H\text{-}bark? = y$ means that I hear something which I interpret as my dog's barking.

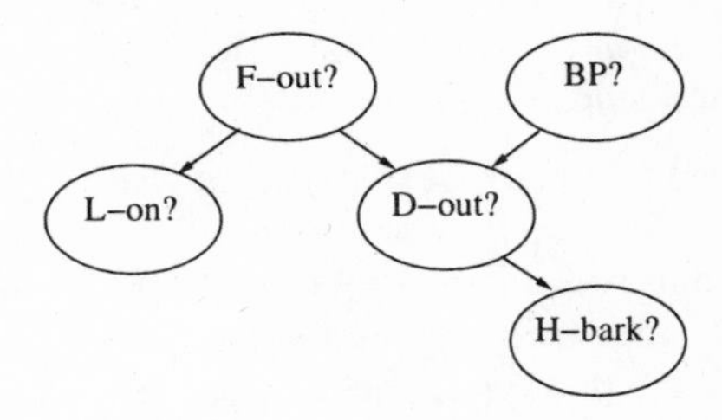

Figure 3.13 The final model for *Family out?*. We have introduced the mediating variable *D-out?* with the additional parent *BP?*.

3.3 Modelling tricks

Much scepticism to Bayesian networks stems from the question "Where do the numbers come from?". As shown in the previous section, they come from many different sources. If you are building a model over a domain where experts actually *do* take decisions based on estimates, why shouldn't you be able to make your Bayesian network estimate at least as well as the experts? You can, for example, use the technique described in Section 2.3.3 to acquire the probabilities from the experts. The acquisition of numbers is of course not without problems, and in this section we give some methods which can help you in this job.

3.3.1 Undirected relations

It may happen that the model must contain dependence relations between variables A, B, C, say; but it is neither desirable nor possible to attach directions on them. (In that case the model is called a *chain graph*. A chain graph is an acyclic graph with both directed and non-directed links, where *acyclic* means that all cycles consist of only non-directed links.) The relation may, for example, be a description of possible configurations. This difficulty may be overcome by using conditional dependence as described in Section 2.2.1 (converging influence).

Let $R(A, B, C)$ describe the relation in numbers from [0,1]. Add a new variable D with two states y and n, and let A, B, C be parents of D (see Fig. 3.14).

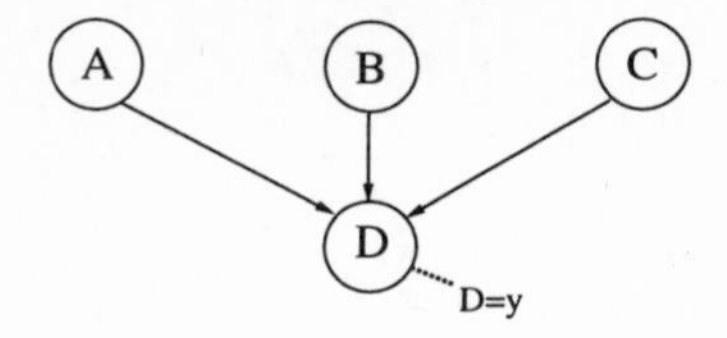

Figure 3.14 A way to introduce undirected relations between A, B and C.

Let $P(D = y \mid A, B, C) = R(A, B, C)$, let $P(D = n \mid A, B, C) = 1 - R(A, B, C)$ and enter the evidence $D = y$.

Example. I have washed two pairs of socks in the washing machine. The washing has been rather hard on them, so they are now difficult to distinguish. However, it is important for me to pair them correctly. To classify the socks I have pattern and colour. A classification model may be like the one in Figure 3.15. The variables S_i have states t_1 and t_2 for the two types, the variables P_i have two pattern types, and the variables C_i have two colour types. The constraint that there are exactly two socks of each type is described in Table 3.10.

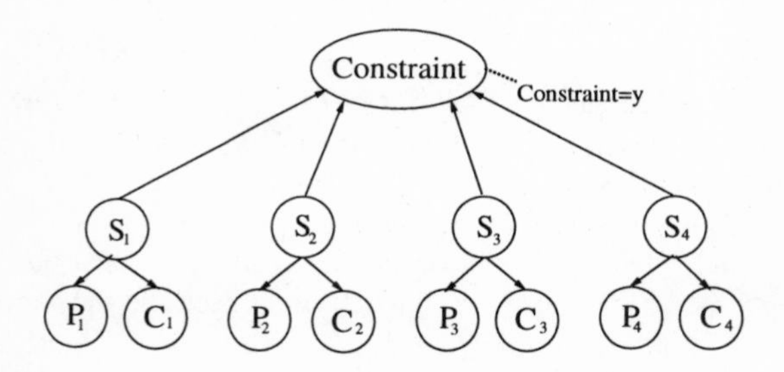

Figure 3.15 A model for classifying pairs of socks.

The situation is more subtle if the relation $R(A, B, C)$ is of a probabilistic nature. If A, B and C have no parents, $R(A, B, C)$ can be a joint probability table. On the other hand, if A has a parent then it is not obvious what $R(A, B, C)$ represents. We shall not deal with this problem but refer the reader to the literature on chain graphs.

3.3.2 Noisy or

When a variable A has several parents you must specify $P(A \mid c^*)$ for each configuration c^* of the parents. If you take the distributions from a data base, the number of cases for each configuration may become too small. Also, the configurations

Table 3.10 The table for $P = P(\text{Constraint} = y \mid S_1, S_2, S_3, S_4)$; t_1 and t_2 are the two states of S_1, S_2, S_3, S_4.

S_1	t_1	t_1	t_1	t_1	t_1	t_1	t_1	t_1	t_2	t_2	t_2	t_2	t_2	t_2	t_2	t_2
S_2	t_1	t_1	t_1	t_1	t_2	t_2	t_2	t_2	t_1	t_1	t_1	t_1	t_2	t_2	t_2	t_2
S_3	t_1	t_1	t_2	t_2	t_1	t_1	t_2	t_2	t_1	t_1	t_2	t_2	t_1	t_1	t_2	t_2
S_4	t_1	t_2	t_1	t_2	t_1	t_2	t_1	t_2	t_1	t_2	t_1	t_2	t_1	t_2	t_1	t_2
P	0	0	0	1	0	1	1	0	0	1	1	0	1	0	0	0

may be too specific for any expert. You may also be in the situation that you have reasonable estimates of $P(A \mid B)$ and $P(A \mid C)$, but you require $P(A \mid B, C)$. Then you should look for assumptions which reduce the amount of distributions to specify.

Table 3.11 Calculation of $P(\textit{D-out?} = y \mid \textit{F-out?}, \textit{BP?})$.

	$\textit{BP?} = y$	$\textit{BP?} = n$
$\textit{F-out?} = y$	$1 - 0.8 \cdot 0.05 \cdot 0.15$	$1 - 0.8 \cdot 0.15$
$\textit{F-out?} = n$	$1 - 0.8 \cdot 0.05$	$1 - 0.8$

Consider in "Family out?" (Section 3.2.4) the conditional probability

$$P(\textit{D-out?} \mid \textit{F-out?}, \textit{BP?}).$$

It was possible to get estimates of $P(\textit{D-out?} \mid \textit{F-out?})$ and $P(\textit{D-out?} \mid \textit{BP?})$, but is there a general way to describe how they combine into $P(\textit{D-out?} \mid \textit{F-out?}, \textit{BP?})$? The following is a way of describing it.

There are three events causing the dog to be outside:

- the "background event" that in 20% of the time the dog is outside "just because";
- *F-out?* which causes the dog to be outside with probability 0.85;
- *BP?* which causes the dog to be outside with probability 0.95.

The above uncertainty can be interpreted in the following way. If any of the causes are present then the dog is outside, unless something has prevented it. In other words, if the family is out then the dog is outside unless they have forgotten to let it out, and there is a 15% chance that they will forget. In the same way there is a 5% chance that some inhibitor prevents the dog from being let out when it has bowel problems and the background event is prevented with probability 0.8.

Now, if we assume that the preventing factors are independent, then the combined probabilities are easy to calculate as one minus the product of the appropriate probabilities for the inhibitors (note that the background event is always a fact). The probabilities are given in Table 3.11.

The construction given above is an example of the simplifying assumption called *noisy or*.

Let $A_1, \ldots, A_n$ be binary variables listing all the causes of the binary variable B. Each event $A_i = y$ causes $B = y$ unless an inhibitor prevents it, and the probability for that is q_i (see Fig. 3.16).

That is, $P(B = n \mid A_i = y) = q_i$. We assume that *all inhibitors are independent.* Then $P(B = n \mid A_1, A_2, \ldots, A_n) = \prod_{j \in Y} q_j$ where Y is the set of indices for variables in the state y. For example

$$
\begin{aligned}
P(B = y \mid A_1 = y, A_2 = y, A_3 = \cdots = A_n = n)& \\
&= 1 - P(B = n \mid A_1 = y, A_2 = y, A_3 = \cdots = A_n = n) \\
&= 1 - q_1 \cdot q_2.
\end{aligned}
$$

By assuming "noisy or", the number of probabilities to estimate grows linearly with the number of parents.

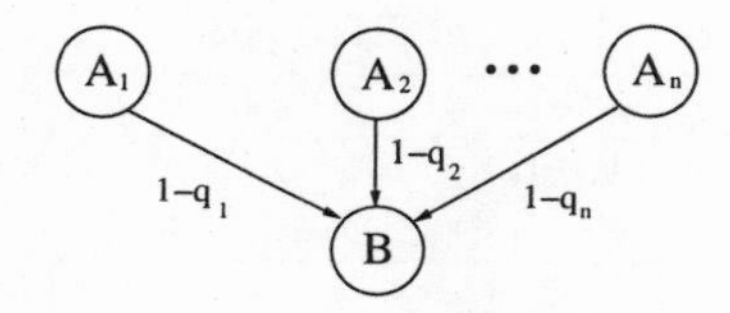

Figure 3.16 The general situation for noisy or. q_i is the probability that the impact of A_i is inhibited.

Note 1. We require $P(B = y \mid A_1 = \cdots = A_n = n)$ to be zero. This may seem to restrict the applicability of the approach. However, as in the example above, if $P(B = y) > 0$ when none of the causing events in the model are on, then introduce a background event which is always on.

Note 2. The complementary construction to noisy or is called *noisy and*. A set of causes shall all be on in order to have an effect. However, the causes have random inhibitors which are mutually independent.

Note 3. The noisy or-gate can be modelled directly without performing the calculations (see Fig. 3.17). This highlights the assumptions behind the noisy or-gate. If a cause is on, then its effect may be prevented by an inhibitor, and the probabilities for the inhibitors to be present are independent.

3.3.3 Causal independence

Let $C_1, \ldots, C_n$ be a list of variables all of which are causes of A. If you wish to specify $P(A \mid C_1, \ldots, C_n)$ you might have a very large knowledge acquisition task ahead of you. Usually it will only be possible to obtain partial specifications like $P(A \mid C_i)$, and in the noisy or case, for example, you have to add some assumptions on how the various impacts on A combine. Often you would have some kind of feeling that the causes act independently on A, but this is not a particularly well defined term.

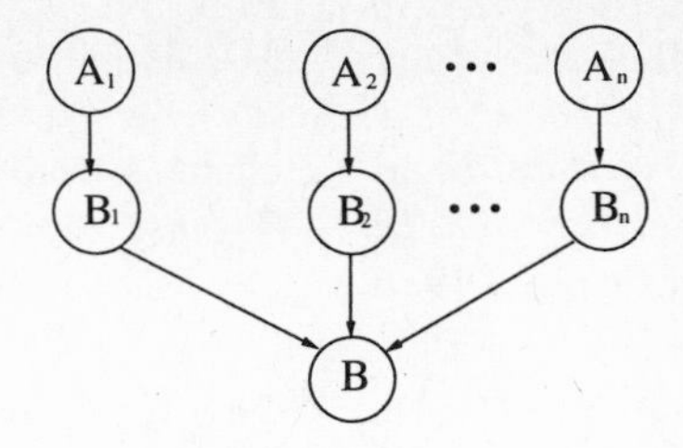

Figure 3.17 Direct modelling of a noisy or-gate. $P(B_i \mid A_i)$ is the original $P(B \mid A_i)$, and $P(B \mid B_1, \ldots, B_n)$ is logical or.

Headache. Headache *(Ha)* may be caused by fever *(Fe)*, hangover *(Ho)*, fibrositis *(Fb)*, brain tumor *(Bt)* and other causes *(Ot)*, and you may choose to soothe it with aspirin *(As)*. (We ignore the effect aspirin has on fever.) Let *Ha* have the states *no, mild, moderate, severe*. The various causes support each other in the effect. If for example $Ho = y$ or $Fb = y$ are present then they may yield a *mild Ha*, but if they are both present then the *Ha* would be *moderate*. Furthermore, if also $As = y$ then *Ha* may drop to *no* or *mild*. Although the various parents of *Ha* combine in a rather involved manner we still have the feeling that the causes' impacts are independent. This kind of independence can be described as follows: if the headache is at level l and we add an extra cause for headache then the result is a headache at level q independently of how the initial state has been caused.

More formally it means the following.

Let $C_1, \ldots, C_n$ be the parents of A. $C_1, \ldots, C_n$ are causally independent if the following holds for each parent configuration $(c_1, \ldots, c_n)$ and all i: if at some time A is in state a and the state of C_i is changed to c_i' then the probability distribution of A afterwards is a function of a, c_i and c_i' alone.

If we add auxiliary variables C_i' and A_i' the condition above is reflected in the conditional independences which can be seen in Figure 3.18.

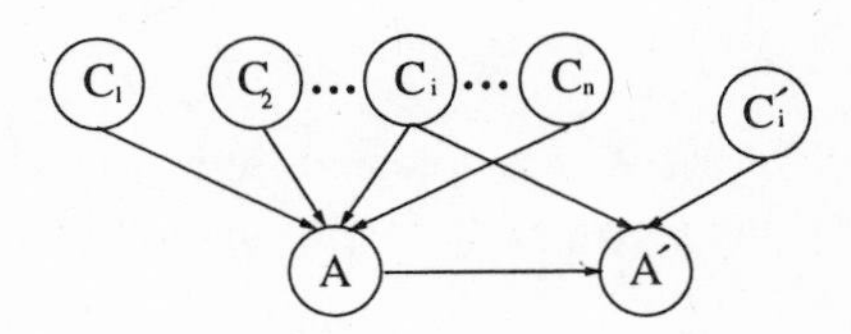

Figure 3.18 A' is independent of $C_1, \ldots, C_{i-1}, C_{i+1}, \ldots, C_n$ given A and C_i.

It can now be seen that aspirin is not causally independent of the causes for headache according to this definition. Suppose that aspirin has been taken without any reason and then a fever arises. Due to the aspirin, the headache will not reach the level it would have reached otherwise. On the other hand, when all present causes for headache are taken into account then the effect of aspirin is only a function of the state of *Ha*.

From the discussion we can conclude that it should be sufficient to estimate the probabilities of the kind $P(Ha - Fe \mid Ha, Fe)$, which shall be estimated for the following rule: if *Ha* is in state *h* and *Fe* changes from *no* to *f* then the probability distribution for headache is $P(Ha - Fe \mid h, f)$. We can then combine these probabilities as shown in Figure 3.19.

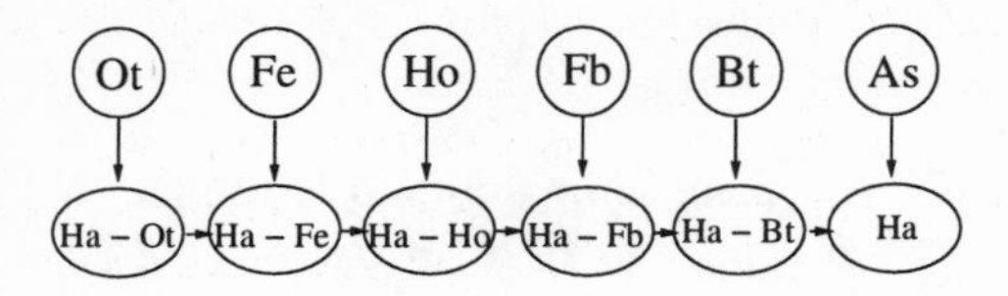

Figure 3.19 A network modelling the multiple cause relation for headache (*Ha*). *Ot*,Other causes; *Fe*, Fever; *Ho*, Hangover; *Fb*, Fibrositis; *Bt*, Brain tumor; *As*, Aspirin.

Note that the chain is started by the background variable *Ot* and the aspirin variable finishes the chain. The order of the intermediate variables is unimportant.

3.3.4 Divorcing

Noisy or, as well as causal independence represent simplifying assumptions to use when the space of parent configurations is too large. Both methods can be seen as special cases of a more general technique called *divorcing*.

The set of parents $A_1, \ldots, A_i$ for B is divorced from the parents $A_{i+1}, \ldots, A_n$ for B by introducing a mediating variable C, making C a child of $A_1, \ldots, A_i$ and a parent of B (see Fig. 3.20).

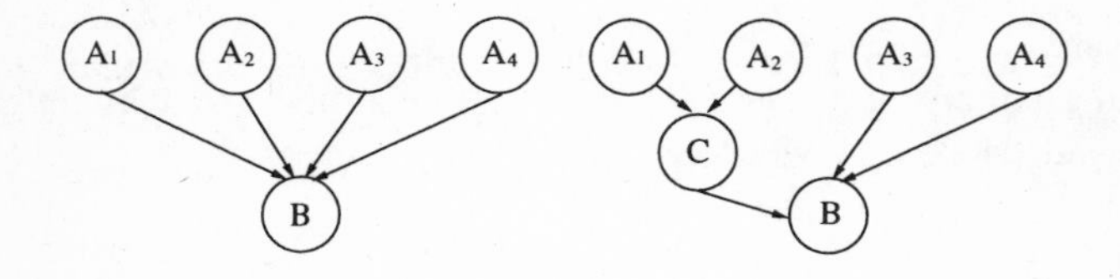

Figure 3.20 A_1 and A_2 are divorced from A_3 and A_4 by introducing the variable C.

If all variables in Figure 3.20 are ternary, you will have to specify 81 distributions before divorcing and only 36 distributions after divorcing. Even if C turns out to require five states, the saving is considerable.

The assumption behind divorcing is the following (with reference to Fig. 3.20).

The set of configurations of (A_1, A_2) can be partitioned into the sets $c_1, \ldots, c_m$ such that whenever two configurations (a_1, a_2) and (a'_1, a'_2) are elements in the same c_i, then $P(B \mid a_1, a_2, A_3, A_4) = P(B \mid a'_1, a'_2, A_3, A_4)$. The divorcing variable then has $c_1, \ldots, c_m$ as states.

Example. To help the bank decide when a customer applies for a mortgage on a house, the customer is asked to fill in a form giving information on various economic and personal matters. The answers are used to estimate the probability that the bank will get their money back.

The information is the following: type of job, yearly income, other financial commitments, number and type of cars in the family, size and age of the house, price of the house, number of previous addresses during the last five years, number of children in the family, number of divorces, and number of children not living in the family.

In principle each slot in the form represents a variable with a causal impact on the variable *money back?*. However, the information can be partitioned into variables describing the economic potentials of the applicant, variables describing the stability of the applicant, and variables describing the security of the mortgage. So, the many parents can be divorced by three variables.

3.4 Learning

By *learning* we understand semi-automatic methods using experience gained to construct or modify a model. Learning can be divided into *qualitative* and *quantitative* learning. Qualitative learning concerns the structure of the model (i.e. the network) and quantitative learning is the specification of conditional probabilities.

There are two types of learning situations. *Batch learning* is the situation where a database of cases is used to establish a model. *Adaptation* is the process of modifying a model successively when new cases are gathered.

3.4.1 Batch learning

In *transmission of symbol strings* (Section 3.2.3) a model was used for the relations between the letters in the words transmitted (see Fig. 3.21), and the table of frequencies (Table 3.7) was used to determine the probabilities for the model. We shall call this model M_{Simp}. However, there may be other models, and we need some means of evaluating these possible models. There are two matters to consider.

- How well can the original table be reconstructed from the model?

- How much space does the model require?

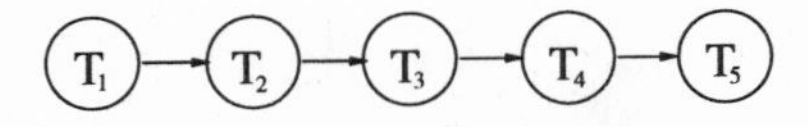

Figure 3.21 The Bayesian network, M_{Simp}, used in Section 3.2.3 for modelling the relation between the letters in the words transmitted

The chain rule (Theorem 2.1) applied to M_{Simp} yields the joint probability table

$$P^*(T_1, T_2, T_3, T_4, T_5) = P(T_1)P(T_2 \mid T_1)P(T_3 \mid T_2)P(T_4 \mid T_3)P(T_5 \mid T_4).$$

The result is shown in Table 3.7. A direct comparison of Table 3.7 and Table 3.7

Table 3.12 $P^*(T_1, T_2, T_3, T_4, T_5)$, the joint probability table determined by M_{Simp}.

	Last 3 letters							
First 2 letters	**aaa**	**aab**	**aba**	**abb**	**baa**	**bab**	**bba**	**bbb**
aa	0.016	0.023	0.018	0.021	0.044	0.067	0.050	0.061
ab	0.030	0.044	0.033	0.041	0.011	0.015	0.012	0.014
ba	0.010	0.016	0.012	0.014	0.029	0.045	0.033	0.041
bb	0.044	0.067	0.059	0.061	0.016	0.023	0.017	0.021

shows some differences, and we may or may not say that M_{Simp} is a sufficiently accurate representation of Table 3.7.

3.4.2 Distance measures

To compare a "true" distribution with an approximation we need a measure of distance between distributions. Let P be a "true" distribution, and let P^* be another distribution over *Word*. Two distance measures are often used.

The *euclidean distance* is

$$\text{Dist}_Q(P, P^*) = \sum_{w \in Word} (P(w) - P^*(w))^2.$$

Cross entropy is

$$\text{Dist}_L(P, P^*) = - \sum_{w \in Word} P(w) \log \frac{P(w)}{P^*(w)}.$$

Note that cross entropy is not symmetric in P and P^*.

The two distance measures have a theoretical foundation, namely *scoring for predictions*: you predict the next word to be transmitted in the form of a probability distribution. When the next word is known, you are penalized by a score such that the penalty is minimal if you had predicted the actual word with probability one, and it is maximal if you assigned probability zero to the actual word.

Let $\text{Act}_w(Word)$ denote the table consisting of zeros except at the place for the actual word, w (where the value is one), and let $P^*(Word)$ be the predicted probabilities. The *quadratic scoring rule* (also called the *Brier scoring rule*) is

$$\text{QS}(w, P^*) = \sum_{x \in Word} (\text{Act}_w(x) - P^*(x))^2 = 1 - 2P^*(w) + \sum_{x \in Word} P^*(x)^2.$$

If you predict a large number of times according to the distribution P^* while the true distribution is P, then the average score will be

$$\begin{aligned} \text{AvQS}(P, P^*) &= \sum_{w \in Word} P(w)\text{QS}(w, P^*) \\ &= 1 - 2 \sum_{w \in Word} P(w)P^*(w) + \sum_{x \in Word} P^*(x)^2. \end{aligned}$$

The quadratic scoring is *strictly proper*. This means that the only distribution yielding a minimal score is the true one (see Exercise 3.17). A strictly proper scoring rule has as a consequence that you are punished if you do not forecast according to your belief (a tempting behaviour if you work with politics). The *distance* between the true distribution P and the estimated distribution P^* is then defined as the difference between the score for the two distributions:

$$\mathrm{Dist}_Q(P, P^*) = \mathrm{AvQS}(P, P^*) - \mathrm{AvQS}(P, P) = \sum_{w \in Word} (P(w) - P^*(w))^2.$$

In our example the true distribution is Table 3.7, and the one used for forecasting is Table 3.7. We get

$$\mathrm{Dist}_Q(P, P^*) = 0.000337.$$

It is more or less up to you to decide how large a distance to accept and fix a threshold. For our tables we would say that the difference between the probabilities should not be larger than 0.004. So, a reasonable threshold would be $32(0.004)^2 = 0.000512$, yielding M_{Simp} acceptable.

Another example of a scoring rule is *logarithmic score:*

$$\mathrm{LS}(w, P^*) = -\log P^*(w).$$

Note that the LS score is extreme if you have assessed a possible event to be impossible. The logarithmic scoring rule is also strictly proper, and the distance measure derived from it is cross entropy.

3.4.3 Search for possible structures

We look for Bayesian networks which represent a joint probability table within an acceptable distance from $P(Word)$; this is not sufficient, though. It may happen that several models are acceptable, and bearing in mind that we look for simple models we also have to take the size of the model into account.

Let M be a Bayesian network with variables U. For each variable A with parents $\mathrm{pa}(A)$ we define $\mathrm{Sp}(A)$ to be the number of entries in $P(A \mid \mathrm{pa}(A))$, and the size is

$$\mathrm{Size}(M) = \sum_{A \in U} \mathrm{Sp}(A).$$

For example, $\mathrm{Size}(M_{\mathrm{Simp}}) = 18$.

To take care of the trade-off between size and distance we define an *acceptance measure:*

$$\mathrm{Acc}(P, M^*) = \mathrm{Size}(M^*) + k\mathrm{Dist}(P, P^*),$$

where P^* is the joint probability table for U determined by M^*, and k is a positive real number.

In the problem with transmission of symbol strings we use $\mathrm{Dist}_Q(P, P^*)$ and choose $k = 10\,000$, and we will work with a distance threshold of 0.0005. The task is to determine an acceptable Bayesian network which minimizes Acc.

In principle, we shall investigate all possible DAGs over the variables T_1, T_2, T_3, T_4, T_5. However, there are too many of them; we therefore add some structure constraints to the models considered. In our case it is natural to say that a directed link from T_i to T_j is only allowed if $i < j$.

We start with the largest model M_{Max} meeting the structure constraint. It is shown in Figure 3.22.

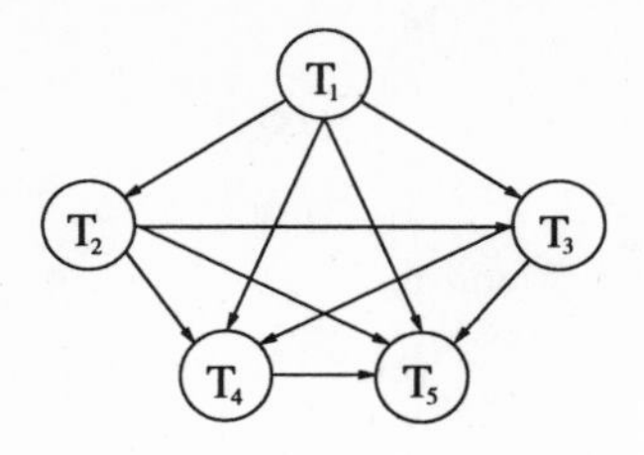

Figure 3.22 The Model M_{Max}, the largest model meeting the constraint that a link from T_i to T_j is only allowed if $i < j$.

Let $P(Word \mid M_{\text{Max}})$ denote the distribution determined by M_{Max}. From the calculation below we see that $P(Word \mid M_{\text{Max}}) = P(Word$:

$$
\begin{aligned}
P(Word) &= P(T_1, T_2, T_3, T_4, T_5) \\
&= P(T_5 \mid T_1, T_2, T_3, T_4)P(T_1, T_2, T_3, T_4) \\
&= P(T_5 \mid T_1, T_2, T_3, T_4)P(T_4 \mid T_1, T_2, T_3)P(T_1, T_2, T_3) \\
&= P(T_5 \mid T_1, T_2, T_3, T_4)P(T_4 \mid T_1, T_2, T_3)P(T_3 \mid T_1, T_2)P(T_2 \mid T_1) \\
&\quad P(T_1) \\
&= P(Word \mid M_{\text{Max}}).
\end{aligned}
$$

Therefore M_{Max} is below the threshold and Acc(P, M_{Max}) is calculated to be 62.

There are 2^{10} DAGs to investigate. This is a heavy task, but if we follow a procedure where we – starting with M_{Max} – successively delete links, we need not remove further when a model is rejected. The result of this search is that the model M_{Min} in Figure 3.23 is the best one.

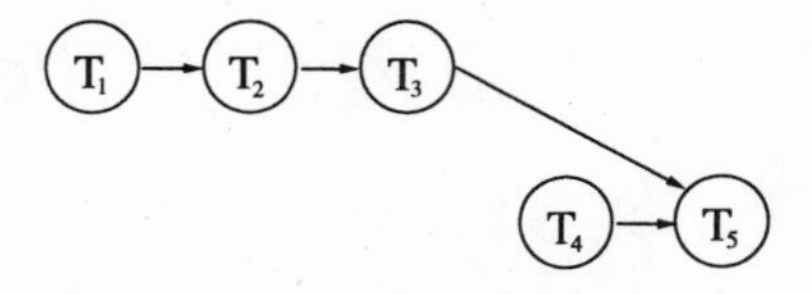

Figure 3.23 M_{Min}. The best model for $P(Word)$.

The tables for M_{Min} are given in Table 13, and the joint probability table determined by M_{Min} is shown in Table 3.14.

We have Acc(P, M_{Min}) = 20.14 and Acc(P, M_{Simp}) = 21.37.

Table 3.13 Conditional probabilities for M_{Min}; $P(T_1) = P(T_4) = (0.5, 0.5)$.

	T_1	
T_2	a	b
a	0.6	0.4
b	0.4	0.6
$P(T_2 \mid T_1)$		

	T_2	
T_3	a	b
a	0.25	0.75
b	0.75	0.25
$P(T_3 \mid T_2)$		

	T_3	
T_4	a	b
a	(0.45, 0.55)	(0.4, 0.6)
b	(0.5, 0.5)	(0.4, 0.6)
$P(T_5 \mid T_3, T_4)$		

3.4.4 Statistical methods

In the considerations so far we assumed the distribution from the database to be the true one, and the task was to find a compact representation approximating it. Usually it is too bold an assumption to consider the database as a true distribution. It is more correct to consider a database as a *sample* from an unknown true distribution. Table 3.7 may, for example, be based on 1,000 words. This means that although M_{Min} is closer to P, then it may still be that P is sampled from M_{Simp}.

Let DB be a database of cases, and let $\mathbb{M}$ be a set of models. What you would try to do is to maximize $P(M \mid \text{DB})$ for $M \in \mathbb{M}$. $P(M \mid \text{DB})$ is a rather awkward probability to calculate. However, Bayes' rule (2.3) can help us:

$$P(M \mid \text{DB}) = \frac{P(\text{DB} \mid M)P(M)}{P(\text{DB})}.$$

Since $P(\text{DB})$ is independent of M it plays no role when determining the maximum. If there is no prior knowledge of the probabilities of the models, we will assume them to be equally likely. Hence the likelihood of M, $P(\text{DB} \mid M)$, can play the same role as distance in the search for a good model. Either you can look for a model of maximal likelihood or you can balance the likelihood with size.

Still, there are many problems. First of all, although there is only a finite number of possible structures, each structure has a continuum of possible tables, and among them we have to find the ones maximizing the probability of DB. Also, very often the database is not a list of cases for which the states of all variables are known. You may have many missing values. It may also happen that you have several databases over different overlapping sets of variables and with different numbers of cases. To make a long story short, the calculation of likelihoods is not a trivial task, and we shall not go into it. In Section 3.7 some references are given.

3.4.5 Adaptation

When a system is at work you repeatedly get new cases, and you would like to learn from these cases. The situation may be that you are pretty certain on the structure of the network, however, the conditional probabilities are dependent on a context which varies from place to place, and you want to build a system which automatically adapts to the particular context in which it is placed. The situation may also be that you have consulted several experts during the construction of the system, and they have not agreed upon the quantitative part of the network. So, the

Table 3.14 $P(Word \mid M_{\mathrm{Min}})$.

	Last 3 letters							
First 2 letters	**aaa**	**aab**	**aba**	**abb**	**baa**	**bab**	**bba**	**bbb**
aa	0.017	0.021	0.019	0.019	0.045	0.068	0.045	0.068
ab	0.034	0.040	0.037	0.038	0.010	0.015	0.010	0.015
ba	0.011	0.014	0.010	0.010	0.031	0.045	0.030	0.045
bb	0.051	0.062	0.057	0.057	0.015	0.023	0.015	0.023

conditional probabilities are uncertain. This type of uncertainty is called *second-order uncertainty*. Second-order uncertainty calls for an automatic way of adapting the conditional probabilities to the real world as it presents itself through the cases.

In Figure 3.24 the variable A is directly influenced by B and C, and the strength is modelled by $P(A \mid B, C)$. The uncertainty in $P(A \mid B, C)$ may be modelled explicitly by introducing an extra parent, T, for A (Fig. 3.24(b)). The variable T can be considered as a type variable, for example types of context or different experts' assessments. To reflect credibility of the experts or frequencies of the context types, a prior distribution $P(T)$ is given.

When a case is entered to the network, the propagation will yield a new distribution $P^*(T)$, and we may say that the change of the distribution for T reflects what has been learnt from the case. $P^*(T)$ can now be used as a new prior distribution. All variables whose table is dependent on the context shall be children of T.

If the uncertainty of the conditional probabilities cannot be modelled explicitly as above, statistical methods can be used. Each entry in a table for a network is a parameter of the model, and the statistical task is to modify the estimates of the parameters gradually with the cases entered. This is an intractable task unless some assumptions on the dependencies between the parameters are added. For the situation above, the dependence is modelled through the T-variable.

Two simplifying assumptions are often used. *Global independence* says that the second-order uncertainty for the various variables is independent.

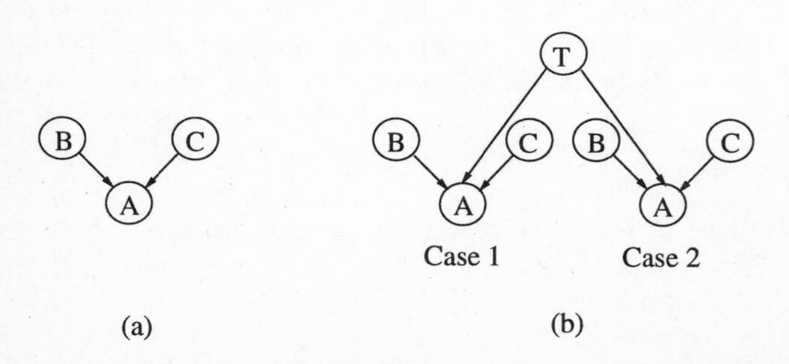

Figure 3.24 Adaptation through a type variable T. The distribution of T is updated by *Case 1* and used in the next case.

Local independence says that the uncertainty of the distributions for different parent configurations are independent. To be more precise, let (a_i, c_j) and (a'_i, c'_j) be different configurations, then the (second-order) uncertainty on $P(A \mid a_i, c_j)$ is independent of the (second-order) uncertainty on $P(A \mid a'_i, c'_j)$.

Consider $P(A \mid B, C)$, and let all variables be ternary. Under the assumption of global and local independence we may now think of $P(A \mid b_i, c_j) = (x_1, x_2, x_3)$ as a distribution established through a number of past cases where (B, C) were in state (b_i, c_j). We can then express our certainty of the distribution by a fictitious sample size s, the larger the sample size the smaller the second-order uncertainty. So, behind the distribution we have a table $(n_1, n_2, n_3) = (sx_1, sx_2, sx_3)$.

When a new case arrives with (B, C) in state (b_i, c_j) and with A, for example, in state a_1, then n_1 and s are counted up by one, yielding a new distribution:

$$(x_1^*, x_2^*, x_3^*) = \left(\frac{n_1 + 1}{s + 1}, \frac{n_2}{s + 1}, \frac{n_3}{s + 1}\right).$$

This scheme only works if the state of A as well as the states of its parents are known. In general we may anticipate that the provided evidence e may leave uncertainty on both the state of A and of its parents.

Let $P(b_i, c_j \mid e) = x$ and $P(A \mid b_i, c_j, e) = (y_1, y_2, y_3)$. A simple approach is to distribute the probability mass x over $P(A \mid b_i, c_j)$ according to the current distribution (y_1, y_2, y_3). Since

$$(y_1 x, y_2 x, y_3 x) = P(A \mid b_i, c_j, e) P(b_i, c_j \mid e) = P(A, b_i, c_j \mid e)$$

we have

$$(n_1^*, n_2^*, n_3^*) = (n_1 + P(a_1, b_i, c_j \mid e), n_2 + P(a_2, b_i, c_j \mid e), n_3 + P(a_3, b_i, c_j \mid e)).$$

Note that the sample size is counted up by $P(b_i, c_j \mid e)$.

This scheme is known as *fractional updating*. Unfortunately, the scheme has a serious drawback, namely that it tends to overestimate the count up of s, thereby overestimating our certainty of the distribution. Assume, for example, that $e = \{B = b_i, C = c_j\}$. Then the case tells us nothing about $P(A \mid b_i, c_j)$, but nevertheless fractional updating will add a count of one to s and take it as a confirmation of the present distribution.

A statistically correct updating is intractable. However, an approximation of it can be performed. It does not have the same drawback as mentioned for fractional updating. Essentially, the distribution is updated as in fractional updating, however, the sample size is modified in a different way. We shall not go into that, but in Section 3.7 some references are given.

3.5 Child

The Great Ormond Street Hospital in London (GOS) specializes in child diseases, and it acts as a regional center for the South-East of England. Whenever a "blue baby" is born in the region, the paediatrician calls the 24 hour telephone service at

GOS, and the clinician at GOS has to come up with a provisional diagnosis based on information provided by the calling paediatrician. The clinician then decides whether or not to transfer the baby to GOS.

To help the clinician, a Bayesian network has been constructed. In fact two networks are constructed, a *subjective* network (*SN*) and a *batch learned* network (*BN*).

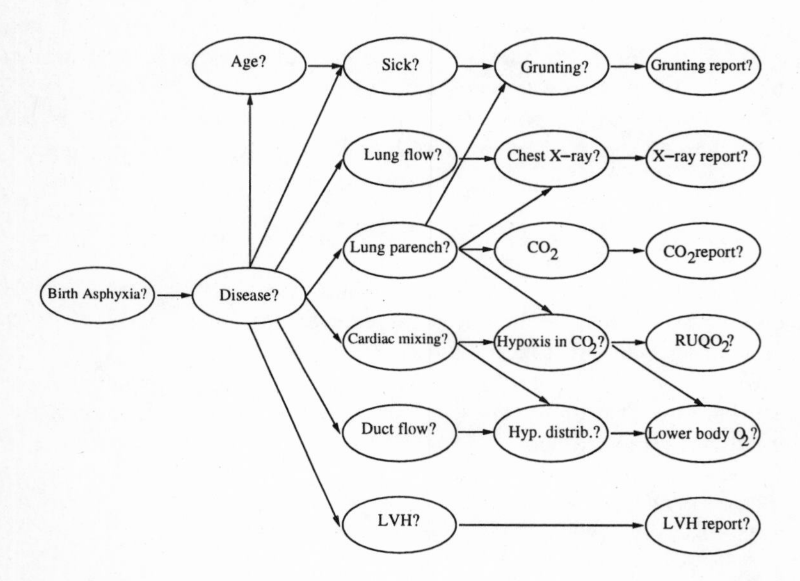

Figure 3.25 The Bayesian network *SN* for congenital heart disease. It is the experts' subjective model. *Disease?* is the only hypothesis variable; *Age?, Birth Asphyxia?, Sick?*, and the rightmost variables are information variables.

SN, which is shown in Figure 3.25, was established through intensive dialogue with the GOS specialists on congenital heart diseases. They provided the model and the conditional probabilities. These probabilities were given together with an estimate of their second-order uncertainty. A set of past cases with known diagnosis was used to adapt the probabilities.

BN was based on 151 cases with established disease, and the resulting model is shown in Figure 3.26. Due to the way batch learning is performed, *BN* can only contain variables from the data base. That is, *BN* does not contain mediating variables.

The two models have been tested in 87 cases (different from the 151 learning cases). The performance of the models was measured through the quadratic scoring rule as well as the logarithmic scoring rule. The conclusion was that both models performed at a level similar to that of the clinicians at GOS, and *SN* performed slightly better than *BN*.

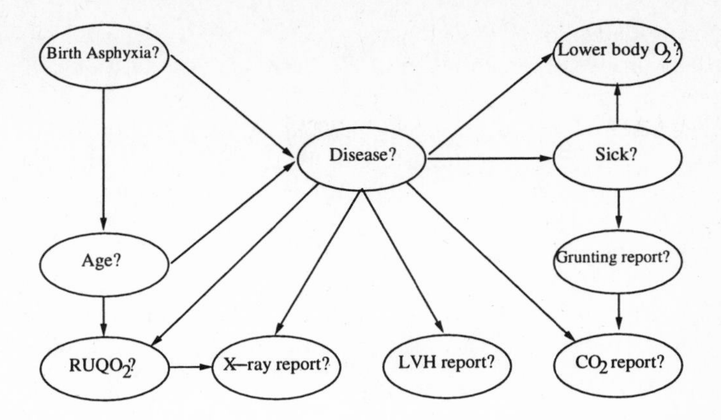

Figure 3.26 The batch learned model for congenital heart disease. The model has no mediating variables.

3.6 Summary

Types of variables when building a Bayesian network model

Hypothesis variables. Variables with a state that is asked for. They are, however, either impossible or too costly to observe.

Information variables. Variables which can be observed.

Mediating variables. Variables introduced for a special purpose. It may be to reflect properly the independence properties in the domain, it may be to facilitate the acquisition of conditional probabilities, it may be to reduce the amount of distributions to acquire for the network, or it may be for other purposes.

Warning: it is tempting to introduce mediating variables in order to have a more refined model of the domain, however, if they do not serve any other purpose you should get rid of them. They jeopardize performance.

Acquiring conditional probabilities

Theoretically well-founded probabilities as well as frequencies and purely subjective estimates can be used for the same network.

If the amount of distributions is too large for a reasonable estimation, simplifying assumptions can reduce it.

Noisy or: Let B have the parents $A_1, \ldots, A_n$ (all variables binary). Suppose that $A_i = y$ causes $B = y$ unless it is inhibited by an inhibitor Q_i which is active with probability q_i. Assume that the inhibitors are independent. Then

$$P(B = n \mid a_1, \ldots, a_n) = \prod_{j \in Y} q_j,$$

where Y is the set of indices for the states y.

Causal independence. Let B have the parents $A_1, \ldots, A_n$. $A_1, \ldots, A_n$ are causally independent with respect to B if the following holds for each configuration $(a_1, \ldots, a_n)$

and all i: if at some time, B is in state b and the state of A_i is changed from a_i to a'_i, then the probability distribution of B afterwards is a function of b, a_i and a'_i alone.

Divorcing. Let B have the parents $A_1, \ldots, A_n$. Assume that the set of configurations of $(A_1, \ldots, A_i)$ can be partitioned into the sets $c_1, \ldots, c_m$ such that whenever two configurations a^* and a_1^* of $(A_1, \ldots, A_n)$ are elements in the same c_i, then

$$P(B \mid a^*, A_{i+1}, \ldots, A_n) = P(B \mid a_1^*, A_{i+1}, \ldots, A_n).$$

Then $A_1, \ldots, A_j$ can be divorced from $A_{i+1}, \ldots, A_n$ by introducing a mediating variable C with states $c_1, \ldots, c_m$, making C a child of $A_1, \ldots, A_i$ and a parent of B.

Batch learning

Let U denote the set of configurations over a universe of variables. Let P be a "true" distribution over U taken from a data base of cases, let M^* be a candidate Bayesian network for P, and let P^* be the distribution determined by M^*.

The euclidean distance.

$$\text{Dist}_Q(P, P^*) = \sum_{x \in U} (P(x) - P^*(x)).$$

The cross entropy distance.

$$\text{Dist}_L(P, P^*) = -\sum_{x \in U} P(x) \log \frac{P(x)}{P^*(x)}.$$

Both distance measures are based on strictly proper scoring rules.

Size measure.

$$\text{Size}(M^*) = \sum_i \text{Sp}(A_i),$$

where $\text{Sp}(A)$ is the number of entries in $P(A \mid \text{pa}(A))$.

Acceptance measure.

$$\text{Acc}(P, M^*) = \text{Size}(M^*) + k\text{Dist}(P, P^*).$$

General search method. Choose a threshold t for $\text{Dist}(P,P^*)$, and a k for $\text{Acc}(P,M^*)$. Among the models with distance to P less than t choose one of minimal $\text{Acc}(P, M^*)$.

Heuristics for searching through the models.

Partition the variables into sets $Z_1, \ldots, Z_m$, and consider only models with links from elements in Z_i to elements in Z_j, with $i \leq j$. Start with the maximal model and delete one link at a time (breadth-first search). Whenever a model M^* with a distance beyond the threshold t is reached, we need not consider submodels of M^*.

Adaptation

Let A be a variable with parents pa(A). Second-order uncertainty is uncertainty on the conditional probability table $P(A \mid \text{pa}(A))$. Adaptation consists of using the incoming cases to reduce second-order uncertainty.

Adaptation through type variables. The second-order uncertainty can be characterized as uncertainty on which table out of $\mathbf{t}_1, \ldots, \mathbf{t}_m$ is the correct one for $P(A \mid \text{pa}(A))$.

Add a type variable T with states $t_1, \ldots, t_m$ and with A as child. The prior probability $P(t_1, \ldots, t_m)$ reflects your belief in the various tables. Let $P(A \mid \text{pa}(A), t_i) = \mathbf{t}_i$.

Whenever a case e has been processed, the probability $P(t_1, \ldots, t_m \mid e)$ is achieved, and it can be used as a prior probability distribution for the next case.

Fractional updating. Assume that the second-order uncertainty obeys both the *global* and the *local independence* requirement. Global independence means that the second-order uncertainty for the various variables is independent. Local independence means that the second-order uncertainty of $P(A \mid a^*)$ and $P(A \mid a_1^*)$ are independent for different configurations a^* and a_1^* of pa(A).

For each parent configuration a^*, choose a fictitious sample size n, expressing the present certainty of $P(A \mid a^*)$. This yields a fictitious sample size $n_a = nP(a \mid a^*)$ for the configuration (a, a^*).

When a case e has been processed it yields $P(a, a^* \mid e)$. Add $P(a, a^* \mid e)$ to n_a. Thereby the sample size is increased by $P(a^* \mid e)$.

Warning: fractional updating reduces the second-order uncertainty too fast.

3.7 Bibliographical notes

The *Family out?* example is borrowed from Charniak (1991). Simple Bayes was used by de dombal et al. (1972). Noisy or was first described by Pearl (1986b); the modelling of causal independence presented here is suggested by Heckerman (1993); divorcing was used in MUNIN Andreassen et al. (1989). Exercise 3.16 is based on Cooper (1990). Chain graphs are treated in depth by Lauritzen (1996).

Learning has a long history in statistics. Edwards & Havranek (1985) introduce a model selection procedure well suited for Bayesian networks. The relation to Bayesian networks was established through work in the early 1990s (Fung & Crawford 1990, Spiegelhalter & Lauritzen 1990a, Dawid & Lauritzen 1993, Cooper & Herskovits 1992). The learning method presented in this chapter is a simplification of the method in BIFROST (Højsgaard & Thiesson 1995). An improved version of fractional updating is presented in Spiegelhalter & Lauritzen (1990), and a further analysis with experiments can be found in Spiegelhalter & Cowell (1992). Buntine (1994) gives an overview of the state of the art of learning with graphical models. Also Heckerman et al. (1994) is a good paper for further studies.

The various versions of Child are documented in Franklin et al. (1989), Franklin et al. (1991), and in Lauritzen et al. (1994).

Table 3.15 Tables for Exercise 3.1.

	$Pr = y$	$Pr = n$
$Ho = y$	0.9	0.01
$Ho = n$	0.1	0.99

	$Ho = y$	$Ho = n$
$BT = y$	0.7	0.1
$BT = n$	0.3	0.9

	$Ho = y$	$Ho = n$
$UT = y$	0.8	0.1
$UT = n$	0.2	0.9

	$Pr = y$	$Pr = n$
$Sc = y$	0.9	0.01
$Sc = n$	0.1	0.99

Exercises

Exercise 3.1 ^{H} Consider the insemination example from Section 3.1.3. Let the probabilities be as Table 3.15. ($Ho = y$ means that hormonal changes have taken place) $P(Pr) = (0.87, 0.13)$.)

(i) What is $P(Pr \mid BT = n, UT = n)$?

(ii) Calculate $P(BT \mid Pr)$ and $P(UT \mid Pr)$ and use them for a simple Bayes model. What is $P(Pr \mid BT = n, UT = n)$ in this model?

Exercise 3.2 Show that the procedure described in Section 3.1.4 is equivalent to updating in the model in Figure 3.6.

Exercise 3.3 Consider the stud farm example in Section 3.2.1 and let the prior probability for aA be 0.005.

(i) Add to the model the frequency 0.001 for mutation of the gene from A to a.

(ii) Construct a model for the situation in part (i), but for a recessive gene borne by the female sex chromosome. (Note that horses with the disease are taken out of production.)

(iii) ^{H} Copy *Stud farm* from the HUGIN diskette and modify the model according to your answers for (i) and (ii).

Exercise 3.4 ^{H} Consider the transmission example from Section 3.2.3.

(i) From Table 3.7 calculate the remaining conditional probabilities for the model in Figure 3.11.

(ii) Implement the model in HUGIN.

(iii) The sequence *baaca* is received. What is the most probable symbols transmitted according to the model in Figure 3.11? What is the most probable word?

(iv) What is the most probable word according to the model in Figure 3.12.

Exercise 3.5 ^{H} Consider the simplified poker game in Section 3.1.2.

(i) Implement the system in HUGIN.

(ii) Extend the system with a facility giving the chances that your hand is better than your opponent's hand.

Exercise 3.6 ^{H} For the BOBLO network in Figure 2.17 the following quantitative relations hold:

- there are four blood-group factors, *F1, F2, F3, F4*, and each factor is either present or absent;
- *F1* is only present if some phenogroup is *f1*;
- *F2* is only absent if both phenogroups are *f2*;
- *F3* is only present if some phenogroup is *f2*;
- *F4* is only absent if both phenogroups are *f3*;
- the prior probabilities for *parental error* are 0.0045 for both being incorrect, 0.0125 for sire incorrect, and 0.0018 for dam incorrect.

(i) Use the information in Section 2.5 to construct a BOBLO network.

(ii) Suppose for each factor that the risk of mistakes by the laboratory is 1 out of 1000. Extend the BOBLO network to incorporate laboratory mistakes.

(iii) For a calf, the stated dam has genotype (*f1, f2*), and the stated sire has (*f1, f3*). The laboratory reports factor 3 to be present and the other factors absent. What are the probabilities for *parental error*?

Exercise 3.7 A new family with two children has moved into the neighbourhood. I notice that one of them is a girl.

(i) Show that the probability that the other child is a boy is $\frac{2}{3}$.

(ii) I now imagine that I ask the girl whether she is the oldest. If she is, then the probability for the second child to be a boy is 0.5. The same holds if she is the youngest child. So I need not ask the question at all. The probability is 0.5. What is wrong?

Exercise 3.8 You are confronted with three doors A, B, and C. Behind exactly one of the doors there is \$10 000. The money is yours if you choose the correct door. After you have made your first choice of door but still not opened it, an official opens another one with nothing behind, and you are allowed to alter your choice. Should you do that?

Exercise 3.9 Extend the model in Figure 3.15 to incorporate constraints on colour and pattern for the same sock.

Exercise 3.10 The *drive* in golf is the first shot when playing a hole. If you drive with a *spoon* (a particular type of golf club) there is a 2% risk of a miss (a bad drive) and out of the good drives $\frac{1}{4}$ have a length of 180 m, $\frac{1}{2}$ are 200 m and $\frac{1}{4}$ have a length of 220 m. You may also use a *driver* (another type of golf club). This will on average increase the length by 10%, but you will also have three times as high a risk of a miss. Now, both wind and the slope of the hole may affect the result of the drive. Wind doubles the risk of a miss, and the length is affected by 10% (longer if the wind is from behind and shorter otherwise). A downhill slope yields 10% longer drives and uphill decreases the length by 10%.

Estimate the probabilities for a miss and length of drive given the various factors.

Table 3.16 Tables for Exercise 3.12.

	No	*Mild*	*Moderate*	*Severe*
No	0.1	0.0	0.0	0.0
Mild	0.8	0.3	0.0	0.0
Moderate	0.1	0.6	0.8	0.0
Severe	0.0	0.1	0.2	1.0

Table (a).

	No	*Mild*	*Moderate*	*Severe*
No	0.3	0.0	0.0	0.0
Mild	0.2	0.4	0.0	0.0
Moderate	0.2	0.2	0.3	0.0
Severe	0.3	0.4	0.7	1.0

Table (b).

	No	*Mild*	*Moderate*	*Severe*
No	1.0	0.7	0.1	0.0
Mild	0.0	0.3	0.7	0.1
Moderate	0.0	0.0	0.2	0.7
Severe	0.0	0.0	0.0	0.2

Table (c).

Exercise 3.11 The *putt* is the (hopefully) last shot on a golf hole. My ball is lying 1 m away from the hole and under normal circumstances I will miss 1 out of 10. However, when it rains I miss 1 out of 7, if it is windy I miss 1 out of 4, if the green is curved I miss 1 out of 3, and if I am putting for a birdie (one under par) I miss 1 out of 2.

Estimate the probabilities for success and failure given the various factors.

Exercise 3.12 ^{H} Consider the headache example in Section 3.3.3. Let $P(Ha) = (0.93, 0.04, 0.02, 0.01)$ and let the tables below describe the effects on headache. Table 3.16(a) describes the effect of fever, hangover and fibrositis. Table 3.16(b) describes the effect of brain tumor, and Table 3.16(c) shows the effect of aspirin.

Use the technique from Figure 3.19 to establish the probabilities for headache given the causes and aspirin.

Exercise 3.13 Show that the model in Figure 3.17 corresponds to the one in Figure 3.16.

Exercise 3.14 ^{H} Show that noisy or may be modelled as described in Figure 3.19 and in Figure 3.20. Apply this to the putting problem of Exercise 3.11, and compare the amount of numbers to specify.

Exercise 3.15 ^{H} Consider the following example of scene interpretation. The image shows a breakfast table for one person, and the task is to determine whether it is a continental or a British breakfast table.

British breakfast is usually composed of tea, bacon and eggs, and toast with marmelade, while continental breakfast consists of coffee, boiled eggs and rolls with jam.

Possible objects. Plate (big or small), cup (tea or coffee), pot (tea or coffee), jar (red or orange contents), cutlery (knife, spoon, fork).

Big plates are confused with small plates with probability 0.1 (and vice versa). Tea cups may be taken for coffee cups with probability 0.3 and coffee cups for tea cups with probability 0.2. Tea pots and coffee pots are confused with probability 0.4, and the colour of the contents of a jar is determined correctly in 95 % of all cases. Knives are taken for spoons with probability 0.05 and for forks with probability 0.1. Spoons are never taken for knives, but for forks with probability 0.25. A fork is recognized as a spoon with probability 0.2, and as a knife with probability 0.1.

Cutlery never come in identical pairs, and if there is a fork, then there is also a knife on the table.

Six objects are identified on the table: a pot, a jar, a plate, a cup, two pieces of cutlery.

Construct a model for interpretation of the scene. It may, for example, be an idea to interpret "usual" as 99 out of 100.

Exercise 3.16 The following relations hold for the Boolean variables A, B, C, D, E and F:

$(A \vee \neg B \vee C) \wedge (B \vee C \vee \neg D) \wedge (\neg C \vee E \vee \neg F) \wedge (\neg A \vee D \vee F) \wedge$
$(A \vee B \vee \neg C) \wedge (\neg B \vee \neg C \vee D) \wedge (C \vee \neg E \vee \neg F) \wedge (A \vee \neg D \vee F)$.

(i) What are the possible configurations?

(ii) We recieve the evidence that A is false and B is true. What are the possible configurations now?

(iii) The *satisfiability problem* for propositional calculus is: given a Boolean expression $\mathbb{E}$ (over n Boolean variables), is there a truth value assignment to the variables which makes $\mathbb{E}$ true?

Show that a method for calculation of probabilities in Bayesian networks yields a method for solving the satisfiability problem for propositional calculus. (Hint: assume that $\mathbb{E}$ is in conjunctive normal form and represent $\mathbb{E}$ as a Bayesian network.)

(iv) Show that a probability calculation in Bayesian networks is NP-hard.

Exercise 3.17 (A proof that the quadratic scoring rule is strictly proper.)

(i) Show that AvQS(P, P^*) is minimal if $P = P^*$. (Hint: use the fact that the function $f(x) = x^2 - 2ax$ is minimal for $x = a$.)

(ii) Show that if AvQS(P, P^*) = AvQS(P, P) then $P = P^*$. (Hint: show that AvQS(P, P^*) − AvQS$(P, P) = \sum_{w \in Word}(P(w) - P^*(w))^2$.)

Chapter 4

Propagation in Bayesian networks

This chapter presents the algorithm used in HUGIN for probability updating in Bayesian networks. The algorithm does not work directly on the Bayesian network, but on a so-called *junction tree* which is a tree of clusters of variables. The clusters are also called *cliques* because they are cliques in a *triangulated graph*, which is a special graph constructed over the network. Each clique holds a table over the configurations of its variables, and HUGIN propagation consists of a series of operations on these tables. The subjects in this chapter are rather mathematical, and the reader interested in the results rather than in the reasoning behind them can jump directly to the summary in Section 4.7, which should give sufficient background for the reading of Chapters 5 and 6.

In Section 4.1 we define the multiplication and division of tables to be used in the algorithm. Section 4.2 gives methods for entering evidence and updating probabilities provided the full joint probability table is available, and in Section 4.3 we give the architecture of the algorithm when the cluster tree is available. Section 4.4 defines the concept junction tree, and we prove the correctness of the algorithm when applied on a junction tree. Section 4.5 is devoted to the construction of a junction tree from the Bayesian network.

The HUGIN algorithm yields the exact updated probabilities, but if you are unlucky, the algorithm will require so much space or time that the task is intractable. In Section 4.6 we present a technique, *stochastic simulation,* which can be used to get approximate probabilities when this happens.

4.1 An algebra of belief tables

Before we treat probability updating, we will introduce more formally the multiplication of belief tables, which we have used implicitly already.

Table 4.1 Multiplication of two tables over $\{A, B\}$; both variables are ternary.

	a_1	a_2	a_3
b_1	x_1	x_2	x_3
b_2	y_1	y_2	y_3
b_3	z_1	z_2	z_3

t

	a_1	a_2	a_3
b_1	x'_1	x'_2	x'_3
b_2	y'_1	y'_2	y'_3
b_3	z'_1	z'_2	z'_3

t′

	a_1	a_2	a_3
b_1	$x_1x'_1$	$x_2x'_2$	$x_3x'_3$
b_2	$y_1y'_1$	$y_2y'_2$	$y_3y'_3$
b_3	$z_1z'_1$	$z_2z'_2$	$z_3z'_3$

$\mathbf{t} \cdot \mathbf{t}'$

4.1.1 Multiplication and division

Let $\mathbf{t}$ and $\mathbf{t}'$ be two tables over the same variables. Then the product $\mathbf{t} \cdot \mathbf{t}'(c^*) = \mathbf{t}(c^*) \cdot \mathbf{t}'(c^*)$ for all configurations c^*.

Table 4.1 gives an example.

If the two tables are over different sets of variables we can also perform a multiplication.

Let $\mathbf{t}_{AB}$ be a table over $\{A, B\}$, and let $\mathbf{t}_{AC}$ be a table over $\{A, C\}$. Then $\mathbf{t}_{AB}$ and $\mathbf{t}_{AC}$ are multiplied by constructing a table $\mathbf{t}_{ABC}$ over $\{A, B, C\}$, and letting $\mathbf{t}_{AB} \cdot \mathbf{t}_{AC}(a, b, c) = \mathbf{t}_{AB}(a, b) \cdot \mathbf{t}_{AC}(a, c)$ for all configurations (a, b, c).

See Table 4.2 for an example.

Table 4.2 Multiplication of $\mathbf{t}_{AB}$ with $\mathbf{t}_{AC}$.

	a_1	a_2
b_1	x_1	x_2
b_2	x_3	x_4

$\mathbf{t}_{AB}$

	a_1	a_2
c_1	y_1	y_2
c_2	y_3	y_4

$\mathbf{t}_{AC}$

	a_1	a_2
b_1	(x_1y_1, x_1y_3)	(x_2y_2, x_2y_4)
b_2	(x_3y_1, x_3y_3)	(x_4y_2, x_4y_4)

$\mathbf{t}_{AB} \cdot \mathbf{t}_{AC}$

Division can be performed in the same way. Only, we have to be careful with zeros. If the denominator table has zero-entries, then the numerator table must have zero at the same places. In that case we put $\frac{0}{0} = 0$.

4.1.2 Marginalization

Let $\mathbf{t}_V$ be a table over V, and let W be a subset of V. A table $\mathbf{t}_W$ over W can be constructed by *marginalization*. For each configuration w^* let $\mathbf{t}_W(w^*)$ be the sum of all $\mathbf{t}_V(v^*)$, where v^* is a configuration of V coinciding with w^*. The notation is

$$\mathbf{t}_W = \sum_{V \setminus W} \mathbf{t}_V.$$

We shall use the following proposition later.

Proposition 4.1 *Let W and V be disjoint sets of variables, and let $\mathbf{t}_W$ and $\mathbf{t}_V$ be tables over W and V. Then*

$$\sum_V (\mathbf{t}_W \cdot \mathbf{t}_V) = \mathbf{t}_W \cdot \sum_V \mathbf{t}_V.$$

That is, tables containing only variables over which you do not marginalize can be taken out of marginalization. See Table 4.3 for an example.

Table 4.3 An example that $\sum_A \mathbf{t}_B \cdot \mathbf{t}_A = \mathbf{t}_B \sum_A \mathbf{t}_A$.

$\mathbf{t}_B$
y_1
y_2
y_3

$\mathbf{t}_A$
x_1
x_2
x_3

$\mathbf{t}_B\mathbf{t}_A$:

	a_1	a_2	a_3
b_1	y_1x_1	y_1x_2	y_1x_3
b_2	y_2x_1	y_2x_2	y_2x_3
b_3	y_3x_1	y_3x_2	y_3x_3

$\sum_A \mathbf{t}_B\mathbf{t}_A$:

b_1	$y_1x_1 + y_1x_2 + y_1x_3$
b_1	$y_2x_1 + y_2x_2 + y_2x_3$
b_1	$y_3x_1 + y_3x_2 + y_3x_3$

$\mathbf{t}_B \sum_A \mathbf{t}_A$:

y_1	
y_2	$(x_1 + x_2 + x_3)$
y_3	

4.2 Probability updating in joint probability tables

Let A be a variable with $P(A) = (x_1, \ldots x_n)$. Assume we get the information e that A can only be in states i and j. This statement says that all states except i and j are impossible, and we have the belief $P(A, e) = (0, \ldots, 0, x_i, 0, \ldots, x_j, 0, \ldots, 0)$. Note that $P(e)$, the prior probability of e, is $x_i + x_j$, the sum of the probabilities of the possible states. To calculate $P(A \mid e)$ we use the fundamental rule:

$$P(A \mid e) = \frac{P(A, e)}{P(e)} = \frac{P(A, e)}{\sum_A P(A, e)}.$$

The way that e is entered can be interpreted as a multiplication of $P(A)$ with the table $\underline{e} = (0, \ldots, 0, 1, 0, \ldots, 0, 1, 0, \ldots, 0)$ resulting in $P(A, e)$.

Definition. Let A be a variable with n states. A *finding* on A is an n-dimensional table of zeros and ones.

Semantically, a finding is a statement that certain states of A are impossible.

Now, let U be a universe of variables, and assume that we have easy access to $P(U)$, the joint probability table. Then, $P(B)$ for any variable B in U is easy to calculate:

$$P(B) = \sum_{U \setminus \{B\}} P(U).$$

Suppose we wish to enter the above finding. Then $P(U, e)$ is the table resulting from $P(U)$ by giving all entries with A in state i or j the value zero and leaving the other entries unchanged. Again, $P(e)$ is the sum of all entries in $P(U, e)$ and

$$P(U \mid e) = \frac{P(U, e)}{P(e)} = \frac{P(U, e)}{\sum_U P(U, e)}.$$

Note that $P(U, e)$ is the product of $P(U)$ and the finding $\underline{e}$. If e consists of several findings $\{f_1, \ldots, f_m\}$ each finding can be entered separately, and $P(U, e)$ is the product of $P(U)$ and the findings $\underline{f}_i$. We can express the considerations above in the following theorem.

Theorem 4.1 *Let U be a universe of variables and let $e = \{f_1, \ldots, f_m\}$. Then*

$$P(U, e) = P(U) \cdot \underline{f}_1 \cdot \cdots \cdot \underline{f}_m \text{ and } P(U, e) = \frac{P(U \mid e)}{P(e)},$$

where

$$P(e) = \sum_U P(U, e).$$

Theorem 4.1 says that if we have access to $P(U)$, then we can enter evidence and perform probability updating. However, even for small sets of variables, the table $P(U)$ is intractably large, and we have to find a smaller representation.

4.3 Cluster trees

As shown in Section 2.3.7 (the chain rule), a Bayesian network over U is a representation of $P(U)$. This means that we can, in principle, calculate $P(U)$ as the product of all conditional probabilities from the network. The question then, is whether we can enter evidence and perform probability updating in Bayesian networks without being forced to calculate $P(U)$. It has turned out to be rather difficult.

Instead we can work with another representation called *cluster trees.*

Definition. A *cluster tree over U* is a tree of clusters of variables from U. The nodes are subsets of U, and the union of all nodes is U. (A tree is an undirected graph without cycles.)

The links are labelled with *separators* which consist of the intersection of the adjacent nodes.

Each node and separator holds a real numbered table over the configurations of its variable set.

In Figure 4.1 we give a cluster tree for the network $M_{\min}$

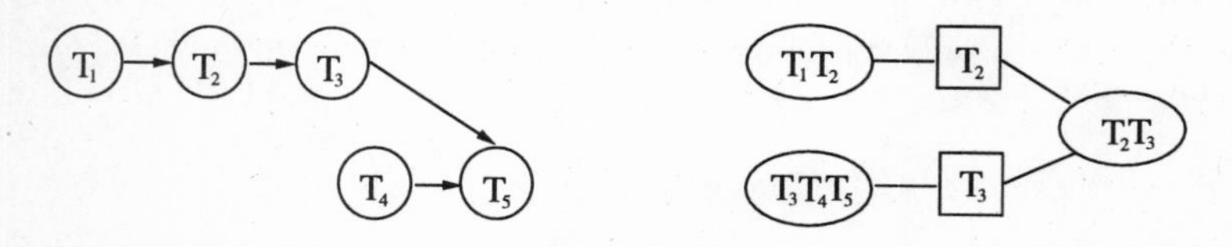

Figure 4.1 The Bayesian network $M_{\min}$ and a corresponding cluster tree. Separators are in square boxes.

Now, let *BN* be a Bayesian network over U. A *Cluster tree corresponding to BN* is constructed in the following way:

- form a family of nodes such that for each variable A with parent set $pa(A)$ there is at least one node V such that $pa(A) \cup \{A\} \subseteq V$;

- organize the nodes as a tree with separators (so far there is no restriction on how you organize the tree);

- give all nodes and separators a table of ones.

- for each variable A choose exactly one node V containing $pa(A) \cup \{A\}$ and multiply $P(A \mid pa(A))$ on Vs table.

Then the product of all node tables in the cluster tree is the product of all conditional probability tables in *BN*, and therefore we have the following theorem.

Theorem 4.2 *Let* BN *be a Bayesian network over* U. *Then any cluster tree corresponding to* BN *is a representation of* $P(U)$, *and* $P(U)$ *is the product of all cluster tables divided by the product of all separator tables.*

Remark. In Theorem 4.2 we divide the product of all cluster tables by the product of all separator tables. This does not do any harm, because the separator tables consist of ones, but the reader may wonder why. The reason is that, when we now start to move the information around in the cluster tree, then the product of all cluster tables divided by all separator tables is invariant, and thereby the tree remains a representation of $P(U)$.

It is easy to insert findings into a cluster tree. Let e be a finding on A. Multiply $\underline{e}$ on the table of any node containing A. Then, by the chain rule and Theorem 4.1 the product of all node tables is $P(U) \cdot \underline{e} = P(U, e)$.

To calculate $P(B, e)$ for an arbitrary variable B is not as easy, and the coming sections are devoted to this problem.

4.3.1 Absorption in cluster trees

We introduce an operation in cluster trees. It has the effect of re-arranging the information stored in the tables.

Definition. Let V and W be neighbours in a cluster tree, let S be their separator, and let $\mathbf{t}_V$, $\mathbf{t}_W$ and $\mathbf{t}_S$ be their tables. The operation *absorption* is the result of the following procedure:

- calculate $\mathbf{t}_S^* = \sum_{V \setminus S} \mathbf{t}_V$;

- give S the table $\mathbf{t}_S^*$;

- give W the table $\mathbf{t}_W^* = \mathbf{t}_W \frac{\mathbf{t}_S^*}{\mathbf{t}_S}$.

We then say that W has *absorbed* from V or that W calibrates to V.

Figure 4.2 W absorps from V. $\mathbf{t}_S^* = \sum_{V\setminus S}\mathbf{t}_V$; $\mathbf{t}_W^* = \mathbf{t}_W \cdot \frac{\mathbf{t}_S^*}{\mathbf{t}_S}$.

Remarks.

(1) The idea behind absorption is that the information which V and W can have in common is the information on S, and this is what W receives from V. If W, V and S hold the same information on S, that is if

$$\sum_{W\setminus S}\mathbf{t}_W = \mathbf{t}_S = \sum_{V\setminus S}\mathbf{t}_V,$$

then absorption does not change anything. We then say that the link is *consistent.* If all links in the cluster tree are consistent we say that the tree is consistent. If a tree is consistent, then absorption does not have any effect at all.

Assume that the link is consistent, but now some evidence changes $\mathbf{t}_V$ to $\mathbf{t}_V^*$. Then after W has absorbed from V, the three tables all hold Vs new information on S:

$$\sum_{W\setminus S}\mathbf{t}_W^* = \sum_{W\setminus S}\mathbf{t}_W\frac{\mathbf{t}_S^*}{\mathbf{t}_S} = \frac{\mathbf{t}_S^*}{\mathbf{t}_S}\sum_{W\setminus S}\mathbf{t}_W = \frac{\mathbf{t}_S^*}{\mathbf{t}_S}\mathbf{t}_S = \mathbf{t}_S^* = \sum_{V\setminus S}\mathbf{t}_V^*.$$

(2) W can only absorb from V through S if $\mathbf{t}_W$ has zeros in the entries corresponding to the zero-entries in $\mathbf{t}_S$. We say that a link in a cluster tree is *supportive* if it allows absorption in both directions, and a cluster tree is supportive if all its links are supportive. Note that the cluster trees constructed in Section 4.2 are supportive since the separator tables have no zero-entries.

Lemma 4.1 *Supportiveness is preserved under absorption.*

Proof. Let W absorb from V through the separator S. Then

$$\mathbf{t}_W^* = \mathbf{t}_W \cdot \frac{\mathbf{t}_S^*}{\mathbf{t}_S},$$

where

$$\mathbf{t}_S^* = \sum_{V\setminus S}\mathbf{t}_V.$$

Then any zero-entry in $\mathbf{t}_S^*$ is also a zero-entry in $\mathbf{t}_W^*$. This clearly also holds for $\mathbf{t}_V$. □

Theorem 4.3 *Let T be a supportive cluster tree. Then the product of all cluster tables divided by the product of all separator tables is invariant under absorption.*

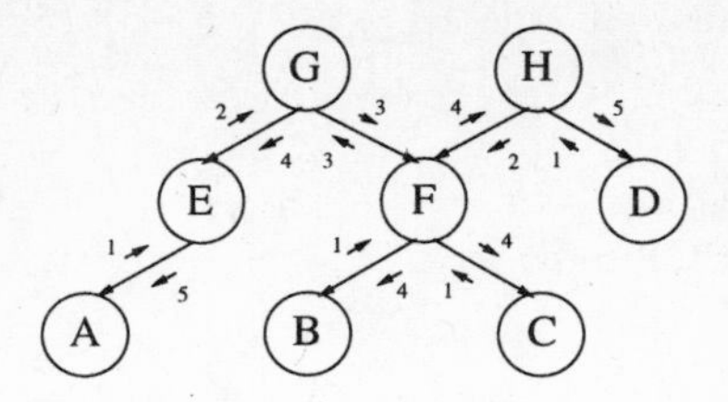

Figure 4.3 Certainty updating through message passing in a cluster tree. The numbers on the links indicate the order in which the messages are passed and in which direction.

Proof. When W absorbs from V through the separator S, only the tables of W and S are changed. Therefore it is enough to prove that the fraction of Ws and Ss table is unchanged. We have

$$\frac{\mathbf{t}_W^*}{\mathbf{t}_S^*} = \frac{\mathbf{t}_W \cdot \frac{\mathbf{t}_S^*}{\mathbf{t}_S}}{\mathbf{t}_S^*} = \frac{\mathbf{t}_W}{\mathbf{t}_S}.$$

Theorem 4.3 ensures that if we start with a Bayesian network over U, construct a corresponding cluster tree T, and then perform a series of absorptions, then T remains a representation of $P(U)$, and $P(U)$ can be calculated as the product of all cluster tables divided by the product of all separator tables.

4.3.2 Message passing in cluster trees

The next question is how many absorptions can we perform, and can they help us in transforming the tables in a cluster tree into a form where it is easy to calculate $P(A)$ for single variables?

We can think of absorptions as messages passed between the nodes in the tree. That is, a node V *sends a message* to its neighbour W when W absorbs from V.

Message passing scheme. A node V can send exactly one message to a neighbour W, and it may only be sent when V has received a message from each of its other neighbours.

Consider, for example, the cluster tree in Figure 4.3. The leaves of the tree (the nodes A, B, C, D) can send to their single neighbour (1). Then E can send to G, and H can send to F (2). Next, G can send to F, and F can send to G (3), F can send to H, B and C, and G can send to E (4). Finally E can send to A and H to D (5). Now each node has sent to all of its neighbours.

As can be seen, the message passing algorithm is not sequential, and a good way of thinking of it is that each variable is busy waiting, eager to send messages. Each time it receives a message it updates its own table and sends a message to the eligible neighbours (if any).

Theorem 4.4 *Let T be a supportive cluster tree, and suppose that messages are passed according to the message passing scheme. Then:*

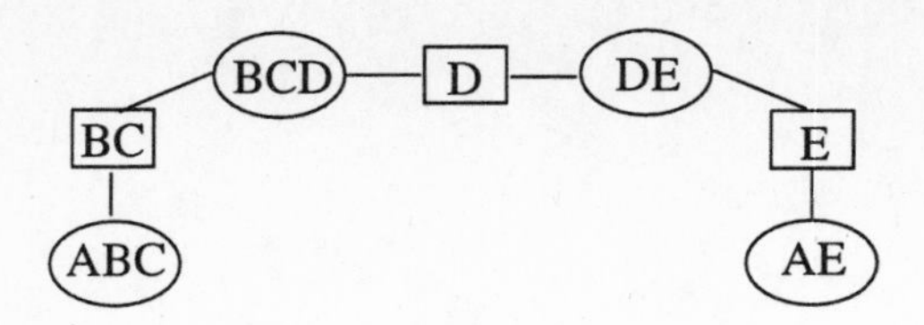

Figure 4.4 A cluster tree over binary variables. All variables except A are in state y. In the node (A, B, C) A is in state y, and in the node (A, E) A is in state n. Though the cluster tree is consistent, the table for $\mathbf{t}_A$ marginalized from $\mathbf{t}_{ABC}$ is different from the marginal taken from $\mathbf{t}_{AE}$.

(i) message passing can continue until a message has been passed in both directions of each link;

(ii) when a message has been passed in both directions of each link then T is consistent.

Proof. (i) Exercise 4.3.

(ii) If T consists of only one node then the theorem is obviously true.

Assume that T has more than one node, and let (V, W) be an arbitrary link with separator S. Let the first message to be passed over (V, W) be from W to V, and let $\mathbf{t}_V$, $\mathbf{t}_S$ and $\mathbf{t}_W$ be the tables before the message is passed.

When the message has been passed we have $\mathbf{t}_S^* = \sum_{W \setminus S} \mathbf{t}_W$. Next, when the message from V and W has to be passed, the tables for S and W have not been changed (W has not received further messages). Let the table for V be $\mathbf{t}_V^{**}$. After message passing we have

$$\mathbf{t}_S^{**} = \sum_{V \setminus S} \mathbf{t}^{**} \quad \textit{and} \quad \mathbf{t}_W^{**} = \mathbf{t}_W \frac{\mathbf{t}_S^{**}}{\mathbf{t}_S^*}.$$

Now

$$\sum_{V \setminus S} \mathbf{t}_V^* = \sum_{V \setminus S} \mathbf{t}_V \frac{\mathbf{t}_S^{**}}{\mathbf{t}_S^*} = \frac{\mathbf{t}_S^{**}}{\mathbf{t}_S^*} \sum_{V \setminus S} \mathbf{t}_V = \frac{\mathbf{t}_S^{**}}{\mathbf{t}_S^*} \mathbf{t}_S^* = \mathbf{t}_S^{**} = \sum_{W \setminus S} \mathbf{t}_W^{**}.$$

Therefore the link is consistent. □

4.4 Junction trees

Let T be a cluster tree over U, let A be a variable in U, and suppose that A is an element of the nodes V and W. If T is consistent we would expect $\sum_{V \setminus \{A\}} \mathbf{t}_V = \sum_{W \setminus \{A\}} \mathbf{t}_W$. Certainly this is so if V and W are neighbours, but otherwise there is no guarantee. See Figure 4.4 for an example.

We say that a consistent cluster tree is *globally consistent* if for any nodes V and W with intersection I we have

$$\sum_{V \setminus I} \mathbf{t}_V = \sum_{W \setminus I} \mathbf{t}_W.$$

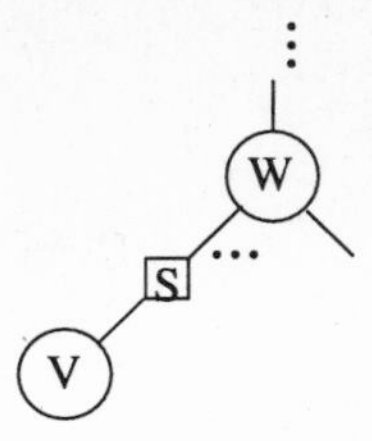

Figure 4.5 V is a leaf of T linked to W and with separator S.

As Figure 4.4 indicates, the reason why consistence does not imply global consistence is that a variable A can be placed in two locations in the tree such that information on A cannot be passed between the two locations. To ensure global consistence we must add a requirement to cluster trees.

Definition. A cluster tree is a *junction tree* if, for each pair of nodes V, W, all nodes on the path between V and W contain the intersection $V \cap W$.

Theorem 4.5 *A consistent junction tree is globally consistent.*

Proof. Exercise 4.7. □

The following theorems will show that if we construct a junction tree corresponding to a Bayesian network, then we have good algorithms for insertion of evidence as well as probability updating. When we construct a cluster tree corresponding to a Bayesian network we have several degrees of freedom, and we shall use them for constructing a junction tree. However, it is not easy. For example, with the clusters in Figure 4.4 it is impossible to construct a tree with the junction tree property. We will leave this problem here, and return to it in Section 4.5.

Theorem 4.6 *Let T be a consistent junction tree over U, and let $\mathbf{t}_U$ be the product of all node tables divided by the product of all separator tables. Let V be a node with table $\mathbf{t}_V$. Then*

$$\mathbf{t}_V = \sum_{U \setminus V} \mathbf{t}_U .$$

Proof. Induction on the number of nodes.

Clearly the theorem holds when T consists of a single node.

Now, assume the theorem to hold for any junction tree with n nodes, and let T be a consistent junction tree with $n+1$ nodes. Let V be a leaf of T linked to W and with separator S (see Fig. 4.5). Let T' be the junction tree resulting from removing V (and S), and let T' have the universe U'. Then

$$\mathbf{t}_U = \mathbf{t}_{U'} \cdot \frac{\mathbf{t}_V}{\mathbf{t}_S} .$$

where $\mathbf{t}_{U'}$ is the product of all node tables in T' divided by the separator tables in T'. Let D be the set of variables $V \setminus S$, and let H be $W \setminus S$. From the junction tree property we have that $D \cap U' = \emptyset$.

Since T is consistent we have

$$\sum_D \mathbf{t}_V = \mathbf{t}_S = \sum_H \mathbf{t}_W.$$

Now

$$\begin{aligned}\sum_D \mathbf{t}_U &= \sum_D \mathbf{t}_{U'} \cdot \frac{\mathbf{t}_V}{\mathbf{t}_S} \\ &= \mathbf{t}_{U'} \cdot \frac{\sum_D \mathbf{t}_V}{\mathbf{t}_S} \\ &= \mathbf{t}_{U'} \cdot \frac{\mathbf{t}_S}{\mathbf{t}_S} \\ &= \mathbf{t}_{U'}.\end{aligned}$$

Therefore, by the induction hypothesis we have

$$\sum_{U \setminus V_i} \mathbf{t}_U = \mathbf{t}_{V_i}$$

for all V_i in T'.

Furthermore,

$$\begin{aligned}\sum_{U \setminus V} \mathbf{t}_U &= \sum_{U' \setminus S} \mathbf{t}_{U'} \cdot \frac{\mathbf{t}_V}{\mathbf{t}_S} \\ &= \frac{\mathbf{t}_V}{\mathbf{t}_S} \cdot \sum_{U' \setminus S} \mathbf{t}_{U'} \\ &= \frac{\mathbf{t}_V}{\mathbf{t}_S} \cdot \sum_{W \setminus S} \mathbf{t}_W \\ &= \frac{\mathbf{t}_V}{\mathbf{t}_S} \cdot \mathbf{t}_S \\ &= \mathbf{t}_V.\end{aligned}$$

□

The considerations above are summarized in the following theorem.

Theorem 4.7 *Let* BN *be a Bayesian network representing* $P(U)$, *and let* T *be a junction tree corresponding to* BN. *After a full round of message passing in* T, *we have for each node* V *and each separator* S *that*

$$\mathbf{t}_V = \sum_{U \setminus V} P(U) = P(V) \textit{ and } \mathbf{t}_S = P(S).$$

Proof. By Theorem 4.2, $P(U)$ is the product of the initial node tables divided by the separator tables. Theorems 4.3 and 4.4 give that after a full round of message passing T is consistent, and $P(U)$ is the product of all node tables divided by all separator tables. Theorems 4.5 and 4.6 yield the conclusion. □

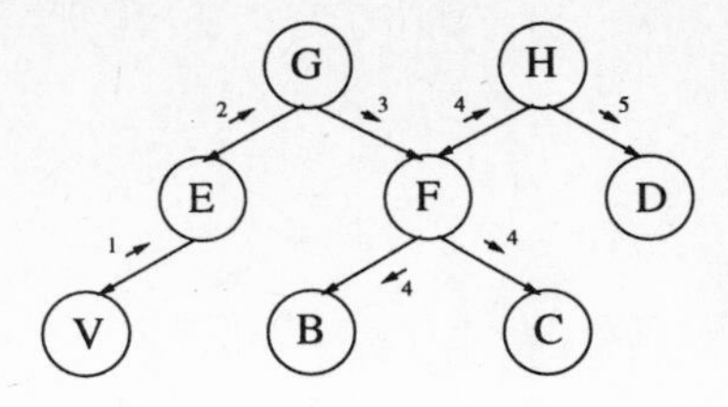

Figure 4.6 The message passing in *DistributeEvidence*(V).

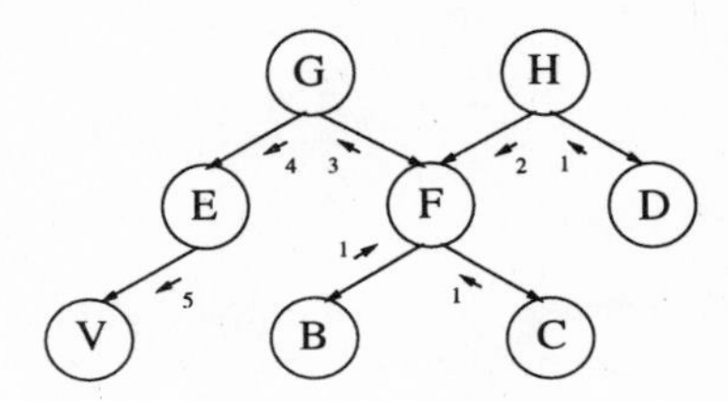

Figure 4.7 The message passing in *CollectEvidence*(V).

Theorem 4.8 *Let* BN *be a Bayesian network representing* $P(U)$*, and let* T *be a junction tree corresponding to* BN. *Let* $e = \{f_1, \ldots, f_m\}$ *be findings on the variables* $\{A_1, \ldots, A_m\}$. *For each* i *find a node containing* A_i *and multiply its table with* $\underline{f}_i$.

Then, after a full round of message passing we have for each node V *and separator* S *that*

$$\mathbf{t}_V = P(V, e) \qquad \mathbf{t}_S = P(S, e) \qquad P(e) = \sum_V \mathbf{t}_V.$$

Proof. Use Theorem 4.1 and proceed as in the proof of Theorem 4.7. □

4.4.1 HUGIN propagation

Assume that we have a consistent junction tree, and now a single node V receives evidence. Then half of the messages can be avoided: V sends messages to all of its neighbours who recursively send messages to all neighbours except the one from which the message came (see Fig. 4.6). We call this algorithm *DistributeEvidence*.

Now, suppose that we are only interested in the certainty of one node, V. Then half of the certainty updating messages can be avoided: V asks all its neighbours to send it a message, and if they are not allowed to do so, they recursively pass the request to all neighbours except the one from which the request came (see Fig. 4.7). We call this algorithm *CollectEvidence*.

The two algorithms *DistributeEvidence* and *CollectEvidence* can be used for a more organized message passing scheme. No matter the amount of evidence entered, take any variable V. Call *CollectEvidence* from V and after that call *DistributeEvidence* from V. The result is that all messages have been passed, and they were passed when permitted (see Fig. 4.8 and Exercise 4.5).

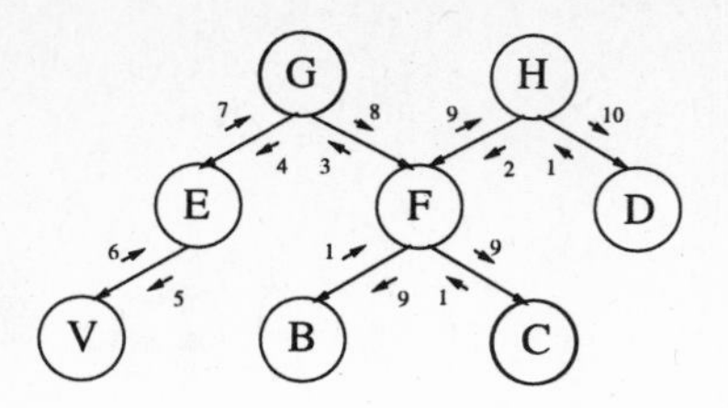

Figure 4.8 Updating through *CollectEvidence*(V) followed by *DistributeEvidence*(V).

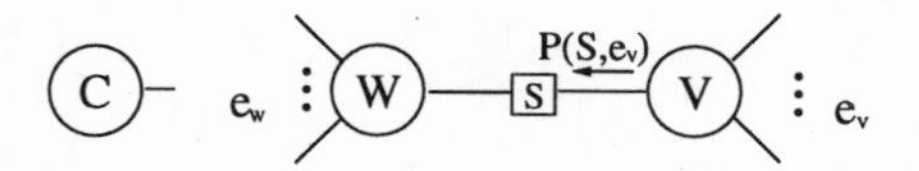

Figure 4.9 Evidence e_V has been entered at the righthand side of S. e_W has been entered at the lefthand side of S. C is used as a root for the propagation.

HUGIN propagation uses corresponding junction trees, and the operations *CollectEvidence* and *DistributeEvidence*. A node Rt in the junction tree is chosen as a root, and whenever a propagation takes place, *CollectEvidence*(Rt) is called followed by a call of *DistributeEvidence*(Rt). When the calls are finished, the tables are *normalized* so that they sum to one.

HUGIN propagation has a nice side effect, namely that it gives access to various probabilities of sets of entered findings.

Let us use Theorem 4.8 to have a closer look at what is actually communicated in the propagation algorithm. The general situation is described in Figure 4.9.

A call of *CollectEvidence*(C) will cause a call of *CollectEvidence*(V), and by Theorem 4.8 this will result in $\mathbf{t}_V^* = P(V, e_V)$. This gives that $P(e_V)$ can be calculated without further propagations. Unfortunately, the situation is not symmetric. In the *DistributeEvidence* phase the message passed from W to S is $P(S, e)$.

4.5 Construction of junction trees

In this section we shall give a method for constructing junction trees for DAGs.

4.5.1 Singly connected DAGs

A DAG is *singly connected* if the graph you get by dropping the directions of the links is a tree (see Fig. 4.10).

For singly connected DAGs it is easy to construct junction trees. For each variable A with $pa(A) \neq \emptyset$ you form the cluster $pa(A) \cup \{A\}$. Between any two clusters with a non-empty intersection you add a link with the intersection as a separator. The resulting graph is called a *junction graph*. All separators consist of a single variable, and if the junction graph has cycles, then all separators on the cycle contain

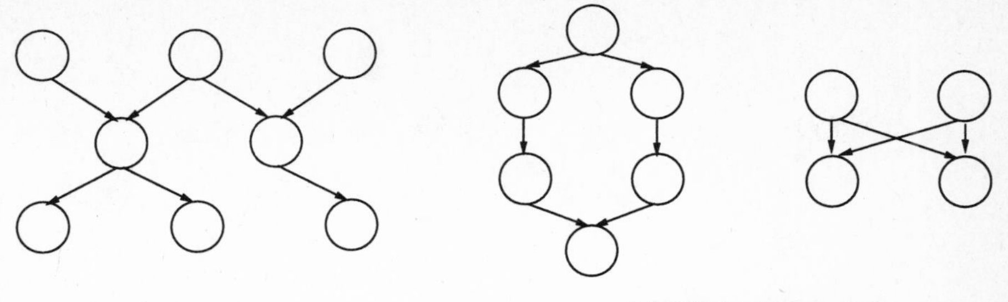

Singly connected Multiply connected

Figure 4.10 Examples of singly connected and multiply connected DAGs.

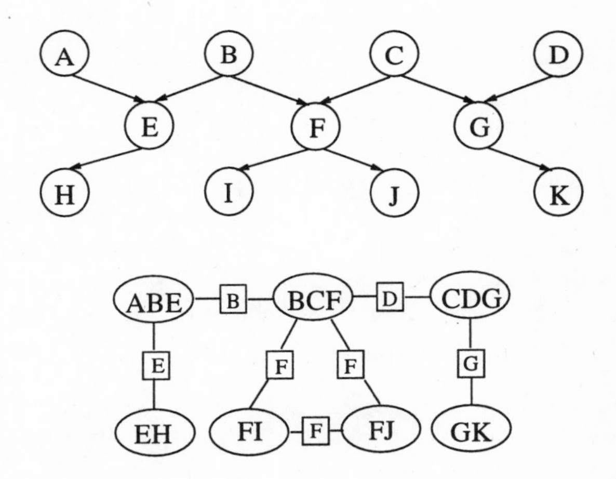

Figure 4.11 A singly connected DAG and its junction graph. By removing any of the links with separator F you get a junction tree.

the same variable. Therefore any of the links can be removed to break the cycle, and by removing links until you have a tree, you get a junction tree (see Fig. 4.11).

We know that when we construct a cluster tree corresponding to a DAG, then for all variables A there must be a cluster V containing $pa(A) \cup \{A\}$. We can illustrate this on a graph by having a link between any pair of variables which must appear in the same cluster. This means that we take the DAG, add a link between any pair of variables with a common child, and drop the directions of the original links. The resulting graph is called the *moral graph.* From the moral graph you can read the clusters to consider, namely the cliques in the graph (maximal sets of variables that are all pairwise linked). In Figure 4.12 we give an example of the construction.

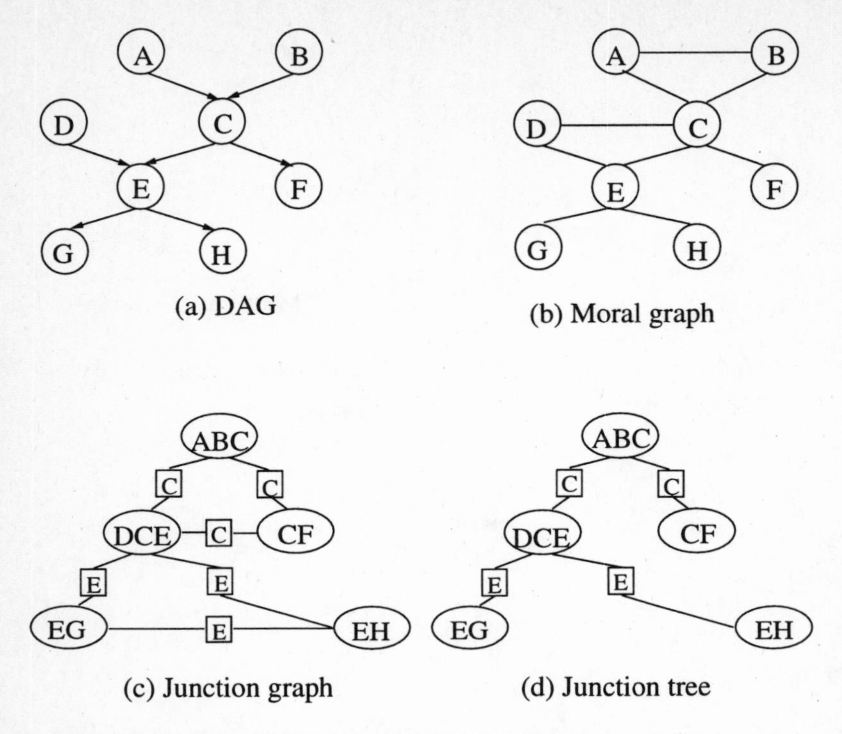

Figure 4.12 Construction of a junction tree for a singly connected DAG.

4.5.2 Coping with cycles

Consider the junction graph in Figure 4.13. The intersection of the two clusters of variables is (AB), and a junction tree is easily found.

Consider the DAG in Figure 4.14(a) with the moral graph in Figure 4.14(b). Sticking to the approach that the clusters are the cliques in the moral graph, we see that if we join A, B and C, then we get a junction tree.

The DAG in Figure 4.15 is more problematic. The cycle in the junction graph cannot be broken.

The propagation problem is that coupled information (on (DE)) is decoupled but meets again under propagation. This can also be seen from the cycle $D - E - C - A - B - D$ in the moral graph. A way to solve the problem is to add so-called *fill-ins* to the moral graph: add a link between C and D and one between B and C. The result is shown in Figure 4.16 together with the resulting junction tree.

The general rule for filling-in the moral graph is that any cycle with more than three variables shall have a chord. In this case the graph is called *triangulated*.

In Figures 4.17 and 4.18 there is another example of the process from DAG to junction tree. Note that without the fill-in $(B - D)$ the cycle $A - B - F - D - A$ does not have a chord.

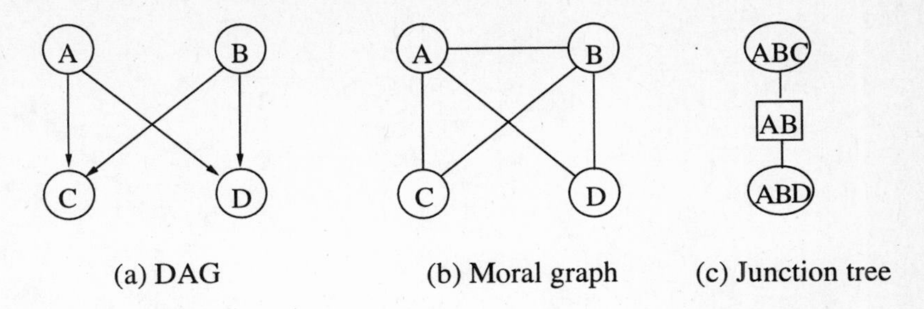

(a) DAG (b) Moral graph (c) Junction tree

Figure 4.13 Construction of a junction tree for a simple multiply connected DAG.

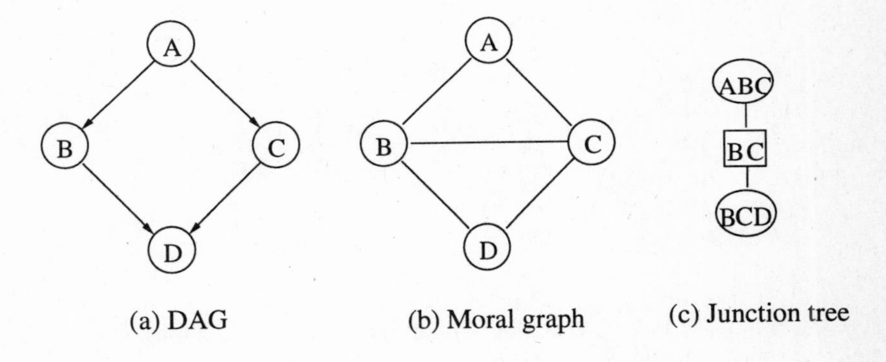

(a) DAG (b) Moral graph (c) Junction tree

Figure 4.14 Another simple DAG with a cycle.

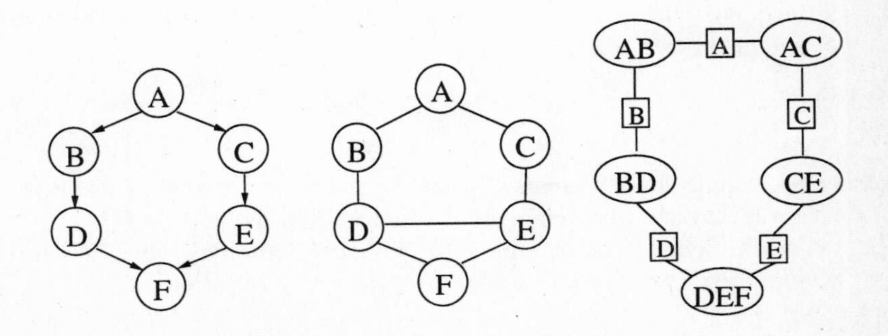

Figure 4.15 A DAG with a large cycle.

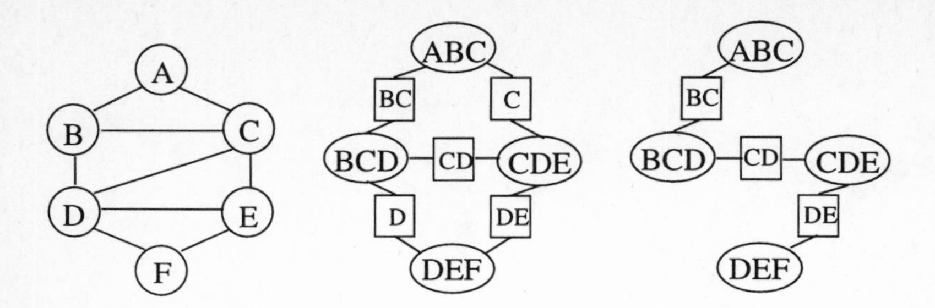

Figure 4.16 The filled-in moral graph from Figure 4.15, the junction graph, and the juction tree resulting from removing the links with separator D and C.

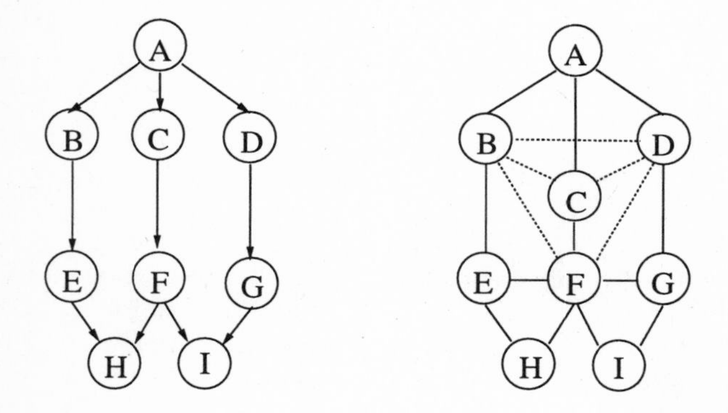

Figure 4.17 A DAG, the moral and triangulated graphs. The fill-ins are indicated by dotted lines.

4.5.3 From DAG to junction tree

In this section we present, without proofs, algorithms for triangulation of graphs and for construction of junction trees from triangulated graphs. Proofs of Theorems 4.9 and 4.10 are given in Appendix A.

Definition. An undirected graph is *triangulated* if any cycle of length > 3 has a chord.

Definition. A node A is *eliminated* by adding links such that all of its neighbours are pairwise linked and then removing A together with its links.

Note that if a node A can be eliminated without adding links, then A cannot be part of a chordless cycle of length > 3.

Theorem 4.9 *A graph is triangulated if and only if all of its nodes can be eliminated one by one without adding any link.*

Theorem 4.9 yields a method for triangulation as well as a test for whether a graph is triangulated. The method consists of eliminating the nodes in some order (adding

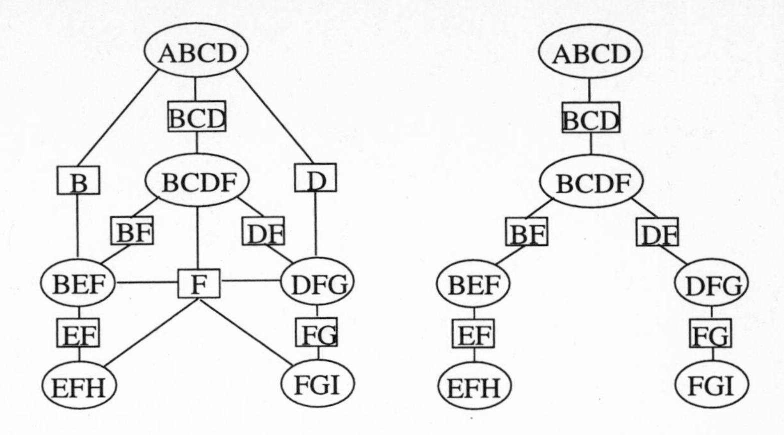

Figure 4.18 The junction graph for the triangulated graph in Figure 4.17 and a junction tree.

links, if necessary) and when this is done the resulting graph is triangulated. An example is given in Figure 4.20.

Note that there are several triangulations of the graph. Intuitively, triangulations with as few fill-ins as possible are preferred. However, optimality is connected to the resulting junction tree and the computational complexity of the propagation algorithm. We shall return to the question of optimality later.

Definition. A *junction graph* for an undirected graph G is an undirected, labelled graph. The nodes are the cliques in G. Every pair of nodes with a non-empty intersection has a link labelled by the intersection.

There is an easy way of identifying the cliques in a triangulated graph G. Let $A_1, \ldots, A_n$ be an elimination sequence for G, and let C_i be the set of variables containing A_i and all its neighbours at the time of elimination (neighbours with higher numbers). Then every clique of G is a C_i for some i.

The reader may check that the cliques of the graphs in Figure 4.20(a) are C_1, C_2, C_3, C_4, and that the cliques of the graph in Figure 4.20(b) are C_1, C_2, C_3.

The junction tree we are aiming at will be a subgraph of the junction graph. Since message passing will be restricted to links in the junction tree we are not allowed to remove a link from the junction graph if thereby some kind of information cannot be passed. If, for example, the clusters U and V have the variable A in common, they have a link with label A. If this link is removed, there shall be another path in the remaining graph through which information on A can be passed from U to V. So, let us recall the following definition.

Definition. A spanning tree of a junction graph is a *junction tree* if it has the property that for each pair of nodes, U, V, all nodes on the path between U and V contain $U \cap V$. (A subtree of a graph is a spanning tree if all nodes of the graph are nodes in it.)

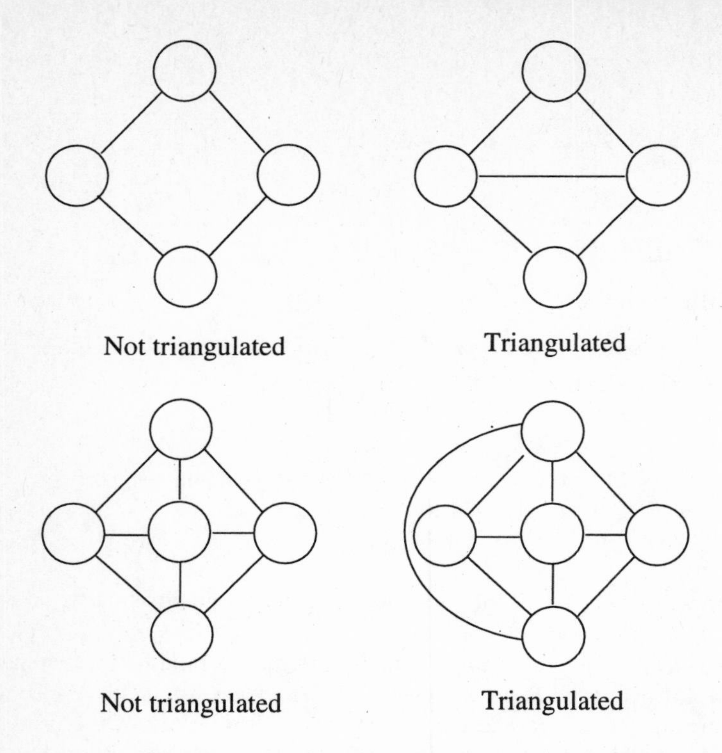

Figure 4.19 Triangulated and not triangulated graphs.

Theorem 4.10 *An undirected graph is triangulated if and only if its junction graph has a junction tree.*

Definition. The *weight* of a link in a junction graph is the number of variables in the label. The weight of a junction tree is the sum of the weights of the labels.

Theorem 4.11 *(Without proof.) A subtree of the junction graph of a triangulated graph is a junction tree if and only if it is a spanning tree of maximal weight.*

Theorem 4.11 provides an easy way of constructing junction trees, namely Kruskal's algorithm: choose successively a link of maximal weight unless it creates a cycle.

There are other ways of constructing junction trees. In particular, if an elimination sequence for the triangulated graph is known, very efficient algorithms exist (see Exercise 4.8). So, if the graph is triangulated then the construction of a junction tree is rather fast.

The only problematic step in the process from DAG to junction tree is the triangulation. Since any elimination sequence will produce a triangulation it may not seem a problem, but for the propagation algorithm it is. In HUGIN propagation the cliques in the junction graph shall have joint probability tables attached to them. The size of the table is the product of the number of states of the variables. So, the size increases exponentially with the size of the clique. A good triangulation,

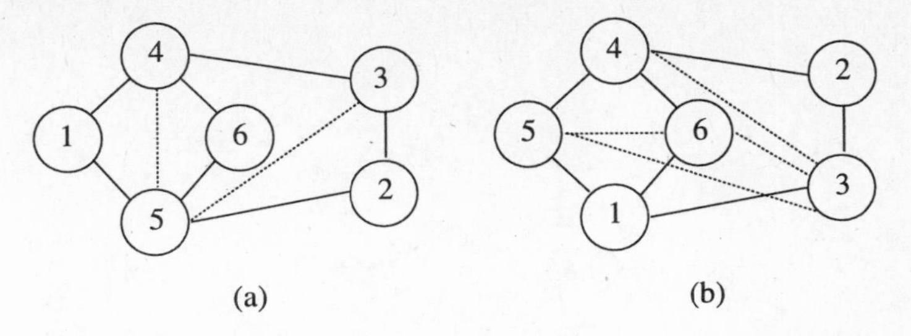

Figure 4.20 Two examples of triangulation through elimination. The numbers on the nodes indicate the elimination order, and the dotted lines are fill-ins.

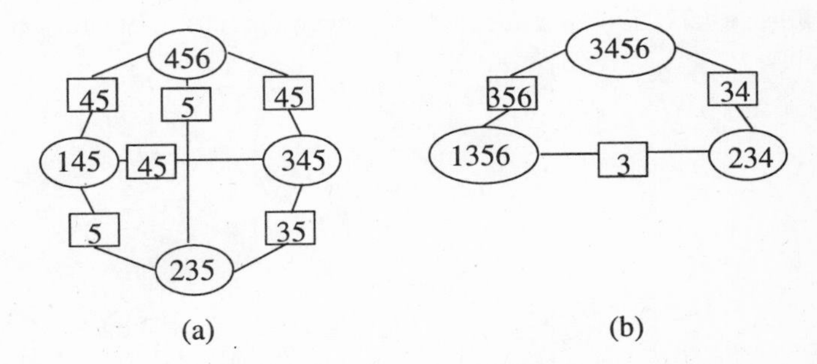

Figure 4.21 Junction graphs for the two triangulated graphs in Figure 4.20.

therefore, is a triangulation yielding small cliques, or to be more precise, yielding small probability tables. The problem of determining an optimal triangulation is NP-complete. However, there is a heuristic algorithm which has proven to give fairly good results. It is a version of the greedy approach: eliminate repeatedly a node not requiring fill-ins and if this is not possible, eliminate a node yielding the smallest table. In Figure 4.23 an example is given.

4.6 Stochastic simulation

The propagation method requires tables for the cliques in the triangulated graph. These cliques may be very large, and it happens that the space requirements cannot be met by the hardware available. In this case an approximate method would be satisfactory.

In this section we shall give a flavour of an approximate method called *stochastic simulation.* The idea behind the simulation is that the causal model is used to simulate the flow of impact. When impact from a set of variables to a variable A is simulated, a random generator is used to decide the state of A.

To illustrate the technique, consider the Bayesian network in Figure 4.24 with the conditional probabilities specified in Table 4.4.

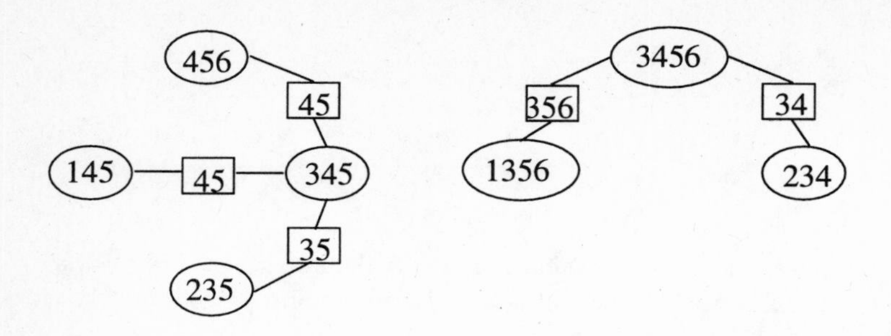

Figure 4.22 Junction trees for the junction graphs in Figure 4.20.

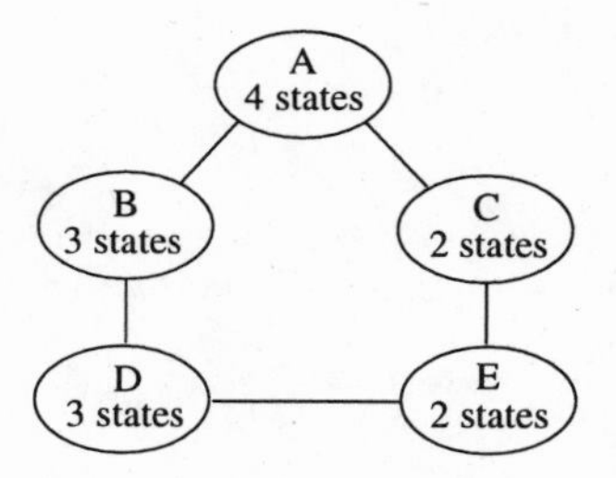

Figure 4.23 A heuristic elimination sequence is E, D (and A, B, C).

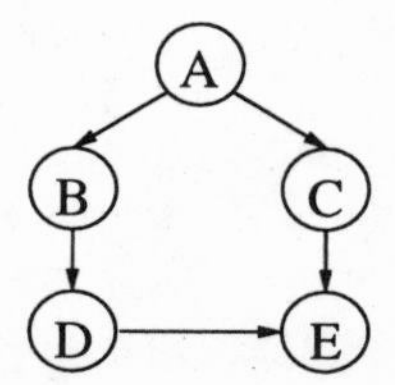

Figure 4.24 An example network. All variables have the states y and n.

Table 4.4 The conditional probabilities for the example network. $P(A) = (0.4, 0.6)$.

	A	
B	y	n
y	0.3	0.8
n	0.7	0.2

$P(B \mid A)$

	A	
C	y	n
y	0.7	0.4
n	0.3	0.6

$P(C \mid A)$

	B	
D	y	n
y	0.5	0.1
n	0.5	0.9

$P(D \mid B)$

	C	
D	y	n
y	(0.9, 0.1)	(0.999, 0.001)
n	(0.999, 0.001)	(0.999, 0.001)

$P(E \mid C, D)$

Table 4.5 A set of 100 configurations of (A, B, C, D, E) sampled from the network in Figure 4.24 and Table 4.4

	CDE							
AB	yyy	yyn	yny	ynn	nyy	nyn	nny	nnn
yy	4	0	5	0	1	0	2	0
yn	2	0	16	0	1	0	8	0
ny	9	1	10	0	14	0	16	0
nn	0	0	4	0	0	0	7	0

The idea now is to draw a random configuration of the variables (A, B, C, D, E), and to do this a sufficient number of times.

A random configuration is selected by successively sampling the states of the variables. First the state of A is sampled. A random generator (with even distribution) is asked to give a real number between zero and one. If the number is less than 0.4 the state is y, if not the state is n. Assume that the result is y. From the conditional probability table $P(B \mid A)$ we have that $P(B \mid y) = (0.3, 0.7)$. The random generator is asked again, and if the number is less than 0.3, the state of B is y. This procedure is repeated to get the state of C, D, and E, and a configuration is determined.

The next configuration is sampled through the same procedure, and the procedure is repeated until m configurations are sampled. In Table 4.5 an example set of configurations is given.

The probability distributions for the variables are calculated by counting in the sample set (see Exercise 4.12). For 39 of the samples in Table 4.5 the first state is y, and this gives an estimated probability $P(A) = (0.39, 0.61)$.

The method above, called *forward sampling*, does not require a triangulation of the network, and it is not necessary to store the sampled configurations (like Table 4.5); it is enough to store the counts for each variable. Whenever a sampled configuration has been determined, the counts of all variables are updated, and the sample can be discarded. This method saves a great deal of space, and each configuration is determined in a time linear to the number of variables. The cost is accuracy and time.

So far only the initial probabilities are calculated. When evidence arrives, it can be handled by simply discarding the configurations which do not conform to it. That is, a new series of stochastic simulations are started, and whenever a state of an observed variable is drawn, you stop simulating if the state drawn is not the observed one.

Unfortunately, this method has a serious drawback. Assume in the example above that the observations for the network are $B = n$ and $E = n$. The probability for $(B = n, E = n)$ is 0.00282. This means that in order to get 100 configurations you should for this tiny example, expect to perform more than 35 000 stochastic simulations.

Methods have been constructed for dealing with this problem. A promising method is called *Gibbs sampling*.

In Gibbs sampling you start with some configuration consistent with the evidence (for example determined by forward sampling), and then you randomly change the state of the variables in causal order. In one sweep through the variables you determine a new configuration, and then you use this configuration for a new sweep, etc.

In the example let $B = n$ and $E = n$ be the evidence, and let the starting configuration be $ynyyn$. Now, calculate the probability of A given the other states of that configuration. That is, $P(A \mid B = n, C = y, D = y, E = n)$. From the network we see that it is sufficient to calculate $P(A \mid B = n, C = y)$. It is easily done by Bayes' rule: it is (0.8, 0.2). We draw a number from the random generator, and let us assume that the number is 0.456 resulting in $A = y$. The next free variable is C. We calculate

$$\begin{aligned} P(C \mid A = y, B = n, D = y, E = n) &= P(C \mid A = y, D = y, E = n) \\ &= (0.996, 0.04). \end{aligned}$$

We draw from the random generator, and assume we keep $C = y$.

In general the calculation goes as follows. Let A be a variable in a Bayesian network *BN*, let $B_1, \ldots, B_n$ be the remaining variables, and let $b^* = (b_1, \ldots, b_n)$ be a configuration of $(B_1, \ldots, B_n)$. Then $P(A, b^*)$ is the product of all conditional tables of *BN* with B_i instantiated to b_i. Therefore $P(A, b^*)$ is proportional to the product of the tables involving A, and $P(A \mid b^*)$ is the result of normalizing this product. Note that the calculation of $P(A \mid b^*)$ is a local task.

Back to the example. The next variable is D. We follow the same procedure and assume that the result is $D = y$. Then the configuration from the first sweep is unaltered, i.e. $ynyyn$.

The next sweep follows the same procedure. Assume the result for A is that the state is changed to n. Then we shall calculate $P(C \mid A = n, D = y, E = n)$, and so forth.

In this way a large sample of configurations consistent with the observations are produced. The question is whether the sample is representative for the probability distribution. It is not always so. It may be that the initial configuration is rather improbable, and therefore the first samples, likewise, are out of the mainstream. Therefore you usually discard the first 5–10% of the samples. It is called *burn-in.*

Another problem is that you may be stuck in certain "areas" of the configurations. Perhaps there is a set of very likely configurations, but in order to reach them from the one you are in, a variable should change to a state which is highly improbable given the remaining configuration (see Exercise 4.13).

A third serious problem is that it may be very hard to find a starting configuration. In fact, it is NP-hard (see Exercise 4.14).

We shall not deal with these problems, but refer the interested reader to the literature.

4.7 Summary of Sections 4.2–4.5

Junction trees

The nodes of a junction tree are sets of variables, they are called *cliques.* Each link is labelled with a *separator* which is the intersection of the adjacent cliques. Each clique and separator holds a real numbered table over the configurations of its variable set.

The junction tree property. For each pair V, W of cliques, all cliques on the path between V and W contain the intersection $V \cap W$.

A junction tree is said to *represent* the Bayesian network BN over the variables U if:

(i) for each variable A, there is a clique containing $pa(A) \cup \{A\}$;

(ii) $P(U)$ is the product of all clique tables divided by all separator tables.

Construction of junction trees

Let BN be a Bayesian network over the variables U.

(i) Construct the *moral graph:* the undirected graph with a link between all variables in $pa(A) \cup \{A\}$ for all A.

(ii) *Triangulate* the moral graph: add links until all cycles consisting of more than three links have a chord.

(iii) The nodes of the junction tree are the cliques of the triangulated graph.

(iv) Connect the cliques of the triangulated graph with links such that a junction tree is constructed.

(v) First give all cliques and separators a table consisting of only ones. Then, for each variable A find a clique containing $pa(A) \cup \{A\}$, and multiply $P(A \mid pa(A))$ on its table.

The resulting junction tree represents BN.

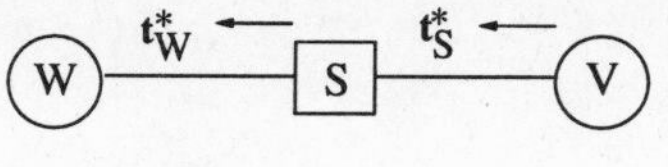

Figure 4.25 W absorps from V. $\mathbf{t}^*_W = \mathbf{t}_W \cdot \frac{\mathbf{t}^*_S}{\mathbf{t}_S}$, $\mathbf{t}^*_S = \sum_{V \setminus S} \mathbf{t}_V$.

Findings

A finding is a statement that some states of a variable are impossible. A finding can be represented as a table of zeros and ones with a zero at the places for impossible states.

A finding on a variable A is entered into a clique V containing A by multiplying Vs table by the table for the finding.

Absorption in junction trees

Definition. Let V and W be neighbours in a junction tree, let S be their separator, and let $\mathbf{t}_V$, $\mathbf{t}_W$ and $\mathbf{t}_S$ be their tables. The operation *absorption* is the result of the following procedure:

- calculate $\mathbf{t}^*_S = \sum_{V \setminus S} \mathbf{t}_V$;
- give S the table $\mathbf{t}^*_S$;
- give W the table $\mathbf{t}^*_W = \mathbf{t}_W \frac{\mathbf{t}^*_S}{\mathbf{t}_S}$.

We then say that W has *absorbed* from V. (See Fig. 4.25.)

HUGIN propagation

An arbitrary clique Rt in the junction tree is chosen as a root. The operation *CollectEvidence* is called in Rt followed by a call of *DistributeEvidence* in Rt.

CollectEvidence(Rt) asks all neighbours to *CollectEvidence* and they proceed down the tree recursively. When all the called neighbours have finished, Rt absorbs from them.

DistributeEvidence(Rt) makes all its neighbours absorb from Rt, and afterwards recursively *DistributeEvidence* to its neighbours (except Rt). See Figure 4.26.

Correctness of HUGIN propagation

Theorem 4.8 Let *BN* be a Bayesian network representing $P(U)$, and let T be a junction tree corresponding to *BN*. Let $e = \{f_1, \ldots, f_m\}$ be findings on the variables $\{A_1, \ldots, A_m\}$. For each i find a node containing A_i and multiply its table with $\underline{f}_i$.

Then, after a full round of message passing we have for each node V and separator S that

$$\mathbf{t}_V = P(V, e) \qquad \mathbf{t}_S = P(S, e) \qquad P(e) = \sum_V \mathbf{t}_V.$$

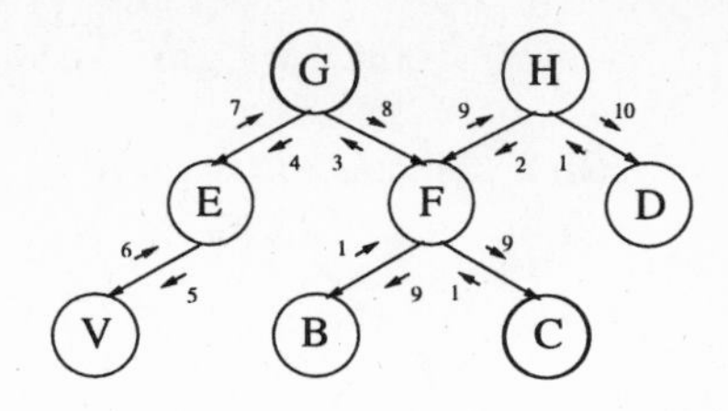

Figure 4.26 Updating through *CollectEvidence*(V) followed by *DistributeEvidence*(V).

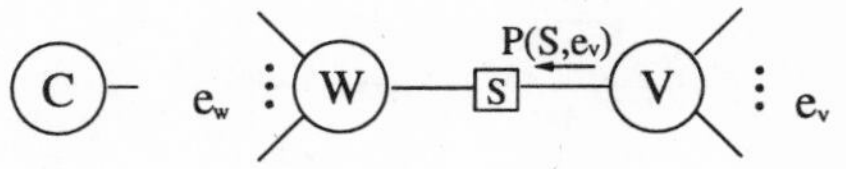

Figure 4.27 Evidence e_V has been entered at the righthand side of S. e_W has been entered at the lefthand side of S. C is used as a root for the propagation.

Side effect of Hugin Propagation

Let Rt be the root for HUGIN propagation, and let W and V be neighbours with separator S. Assume that W is closer to Rt than V. Then S divides the entered evidence in e_V and e_W (see Fig. 4.27).

A call of *CollectEvidence*(Rt) results in the table $P(S, e_V)$ being communicated from V to S. By marginalization you can calculate $P(e_V)$.

4.8 Bibliographical notes

A version of probability updating in singly connected DAGs through message passing was presented by Kim & Pearl (1983). HUGIN propagation was proposed by Jensen et al. (1990). It is a modification of an algorithm proposed by Lauritzen & Spiegelhalter (1988). Similar methods were used for pedigree analysis by Cannings et al. (1978). Shafer & Shenoy (1990) propose a different message-passing method for junction trees. Other propagation methods for multiply connected DAGs exist, e.g. arch reversal proposed by Shachter (1986) or conditioning proposed by Pearl (1986a).

The concepts of triangulated graphs and junction trees have been discovered and rediscovered with various names. In Bertele & Brioschi (1972) they are used for dynamic programming, and Beeri et al. (1983) use them for data base management. A good reference on triangulated graphs is Golumbic (1980). Tarjan & Yannakakis (1984) gives various triangulation methods and very efficient methods for testing whether a graph is triangulated. Jensen & Jensen (1994) contains a proof of Theorem 4.10 together with a method for constructing optimal junction trees from triangulated graphs.

Forward sampling was proposed by Henrion (1988). Gibbs sampling was originally introduced for image restoration by Geman & Geman (1984). Further readings on sampling methods could be Geyer (1992), Fung & Favero (1994), and Jensen et al. (1995). Gilks et al. (1994) have developed a system, BUGS, for Gibbs sampling in Bayesian networks.

Exercises

Exercise 4.1 For Table 4.6, calculate $\mathbf{t}_V \mathbf{t}_W$ and $\frac{\mathbf{t}_W}{\mathbf{t}_V}$.

Table 4.6 Table for Exercise 4.1.

	a_1	a_2	a_3
b_1	1	2	3
b_2	3	2	1

$\mathbf{t}_V$

	c_1	c_2	c_3
b_1	6	12	24
b_2	18	6	12

$\mathbf{t}_W$.

Exercise 4.2 For the universe U over the ternary variables (A, B, C) with the joint probability Table 4.7 we get the findings f_1: "A is in state a_1", and f_2: "C is in state c_1 or c_3".

Table 4.7 Table for Exercise 4.2.

	a_1	a_2	a_3
b_1	(2,4,3)	(1,4,8)	(5,0,7)
b_2	(5,10,4)	(2,3,3)	(1,5,4)
b_3	(1,5,6)	(3,3,3)	(0,6,2)

$P(A, B, C)$ multiplied by ten.

Calculate $P(B \mid f_1, f_2)$, $P(C \mid f_1, f_2)$, $P(f_1)$, $P(f_2)$ and $P(f_1, f_2)$.

Exercise 4.3 Prove that the anarchistic message passing algorithm formulated in Section 4.3.2 never runs into a deadlock: as long as there are unused message channels at least one variable can send a message. (Hint. Induction on the number of nodes and the fact that any sending sequence must start with a leaf sending.)

Exercise 4.4 Let B be independent of C given A, and let $P(A, B)$ and $P(A, C)$ be consistent. What is $P(A, B, C)$?

Exercise 4.5 Prove that a call of *CollectEvidence* in any node followed by a call of *DistributeEvidence* in the same node will result in a full propagation (all messages passed and passed when permitted).

Exercise 4.6 Construct the moral graph and a junction tree for the singly connected DAG below.

Exercise 4.7 Show that a consistent junction tree is globally consistent.

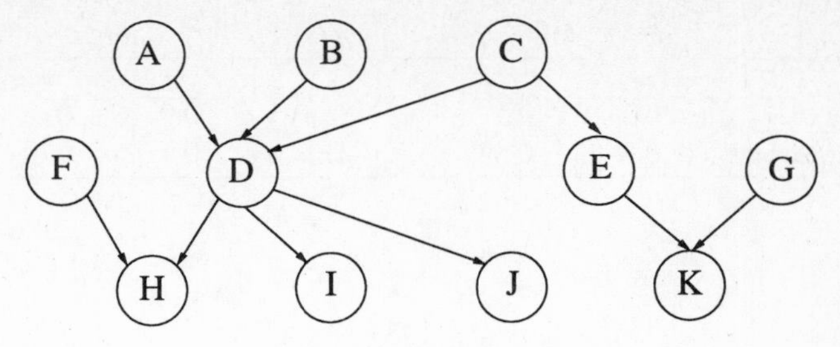

Figure for Exercise 4.6.

Exercise 4.8 (Construction of a junction tree from an elimination sequence.)
G is a triangulated graph over U, and $A_1, \ldots, A_n$ is an elimination sequence of U. C_i is the set of variables containing A_i and all its neighbours at the time of elimination.

(i) Show that each clique of G is a C_i for some i.

(ii) Show that for all $i < n$ there is a $j > i$ such that $C_i \setminus \{A_i\} \subseteq C_j$.

(iii) Assume that C_i and C_j are cliques ($i < j$) such that $C_i \setminus \{A_i\} \subseteq C_j$. Show that there exists a junction tree for G with the link (C_i, C_j).

(iv) Use (ii) and (iii) to construct a junction tree for the graph in Figure 4.20(a).

Exercise 4.9 (i) Construct a junction tree for the DAG given below, by using the elimination order F, J, D, B, A, I, K, E.

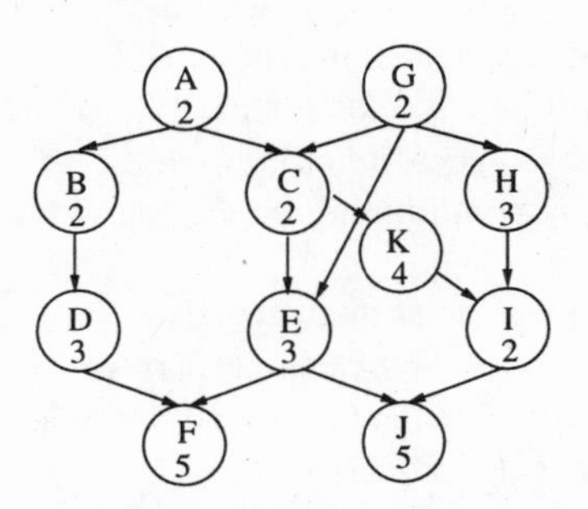

Figure for Exercise 4.9

(ii) The numbers inside the nodes indicate the number of states. Use the procedure from the end of Section 4.5 to construct a junction tree.

Exercise 4.10 (i) For the DAG given below, compute $P(A, B, C)$, when $P(A) = (0.3, 0.7)$ (see Figure and Table 4.8 for Exercise 4.10(i)).

(ii) The DAG is extended as shown in the Figure and Table 4.9 for Exercise 4.10(ii). Calculate $P(B, C, D)$.

Table 4.8 Table for Exercise 4.10(i).

	$A = y$	$A = n$
$B = y$	0.2	0.5
$B = n$	0.8	0.5

$P(B \mid A)$

	$A = y$	$A = n$
$C = y$	0.9	0.4
$C = n$	0.1	0.6

$P(C \mid A)$

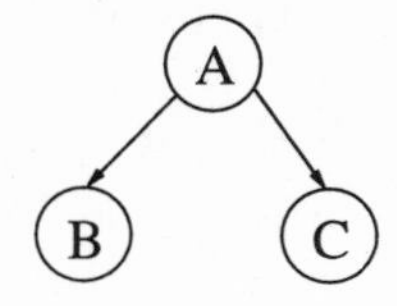

Figure for Exercise 4.10(i).

(iv) We are told that $A = y$ and $D = n$. What is $P(B)$?
(v) Initially, what was $P(A = y, D = n)$?

Exercise 4.11 (Conditioning.) Propagation methods for singly connected DAGs have existed for a long time. A propagation method for multiply connected DAGs consists of reducing a DAG to a set of singly connected DAGs.

(i) Consider the DAG (a) below with $P(A)$, $P(B \mid A)$, $P(C \mid A)$ and $P(D \mid B, C)$ given. Assume that $A = a$. Show that the DAG is reduced to the DAG (b) with $P(B \mid a)$, $P(C \mid a)$, and $P(D \mid B, C)$ given.

(ii) Show that $P(D, a) = P(D \mid b, c)P(B \mid a)P(C \mid a)$.

(iii) Assume that for all states a of A we have a reduced DAG as in (i) . Let evidence e be entered and propagated in all the reduced DAGs, yielding $P(B, e \mid a)$, $P(C, e \mid a)$, $P(D, e \mid a)$ for all a. Calculate $P(B, e)$ and $P(A, e)$.

The procedure above is called *conditioning on A*.

(iv) Reduce the DAG by conditioning on B. Show that the tables are $P(A \mid b)$, $P(C \mid A)$ and $P(D \mid C, b)$.

(v) Show that conditioning on D does not result in a singly connected DAG.

Conditioning over several variables can be performed stepwise.

(vi) Determine a minimal set of conditioning variables for the DAG given below to reduce it to singly connected DAGs.

(vii) The numbers attached to the variables indicate the number of states. Determine a conditioning resulting in a minimal number of singly connected DAGs.

Table 4.9 Table for Exercise 4.10(ii).

	$B = y$	$B = n$
$C = y$	(0, 1)	(0.7, 0.3)
$C = n$	(0.4, 0.6)	(0.5, 0.5)

$P(D \mid B, C)$

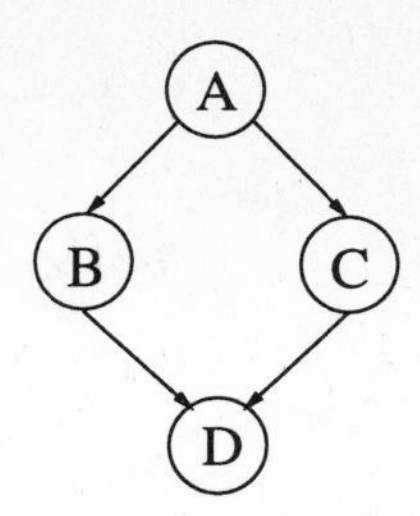

Figure for Exercise 4.10(ii).

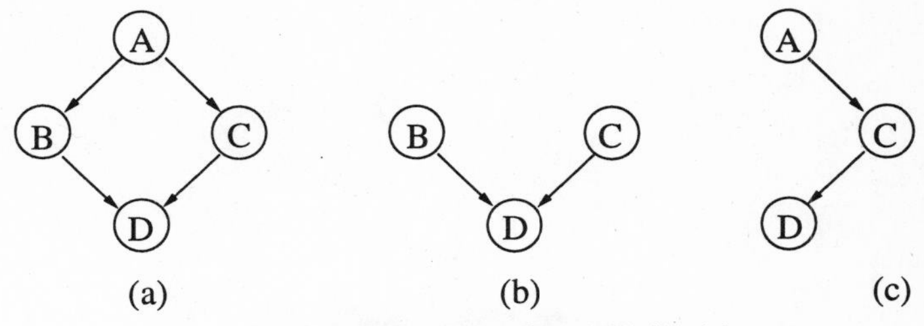

Figure for Exercise 4.11(i)–(v).

Exercise 4.12 Calculate the marginals from the sample in Table 4.5, and compare the result with the exact marginals.

Exercise 4.13 The binary variables A and B are parents of the binary variable C. $P(A) = P(B) = (0.5, 0.5)$, and the conditional probability table is an *exclusive or table*: $C = y$ if and only if exactly one of A and B is in the state y.

Show that Gibbs sampling on this structure will give either $P(C = y) = 1$ or $P(C = n) = 1$.

Exercise 4.14 Given a Bayesian network over U with evidence e entered, show that it is NP-hard to find a configuration U^* such that $P(U^*, e) > 0$.
(Hint. Look at Exercise 3.16.)

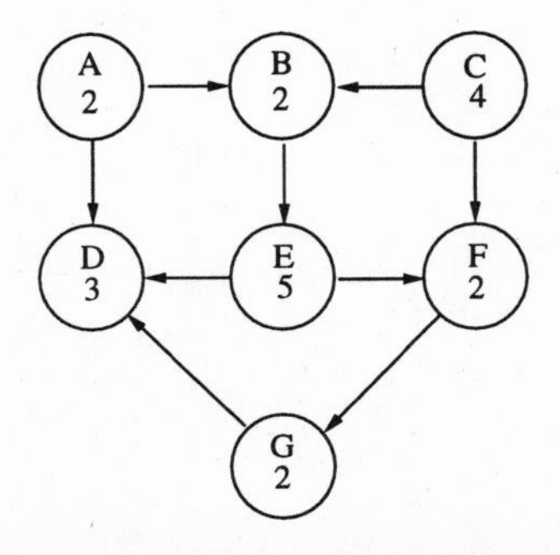

Figure for Exercise 4.11(vi)–(vii).

Chapter 5

Use of Bayesian network models

The main reason for building a Bayesian network is to estimate the state of certain variables given some evidence. In Chapter 4 we gave a method which made it easy to access $P(A \mid e)$ for any variable A. However, this may not be sufficient. It may be crucial to establish the joint probability for a set of variables. Section 5.1 gives a general method for calculating $P(X \mid e)$ for any set X of variables.

Another typical request is to ask for the most probable configuration. We give a method for this in Section 5.2. Section 5.3 deals with methods for analyzing whether the evidence entered to the network is coherent – for example to trace flawed data.

A very important tool for a decision support system is *explanation*: a tool to explain to the user how the system came to its conclusions. A part of explanation is *sensitivity analysis*: how sensitive is the conclusion to (small) changes in the evidence? Which parts of the evidence are crucial and/or sufficient for the conclusion? This is the subject of Section 5.4. Also in Section 5.4 another propagation method, *cautious propagation,* is introduced. Cautious propagation is less efficient than HUGIN propagation. However, it gives as a side effect a very good support for sensitivity analysis.

The last type of request presented in this chapter is *data request*. You are in the middle of a case, and you have various sources for further evidence (each with a cost): which source shall you consult?

Finally we give a short description of the system *Pathfinder*.

5.1 Joint probabilities

We start with some examples.

Transmission of symbol strings. Consider the example in Section 3.2.3 and the model in Figure 3.11. Assume that the sequence *baaca* is received.

(i) What is the probability that the transmitted word is **baaba**?

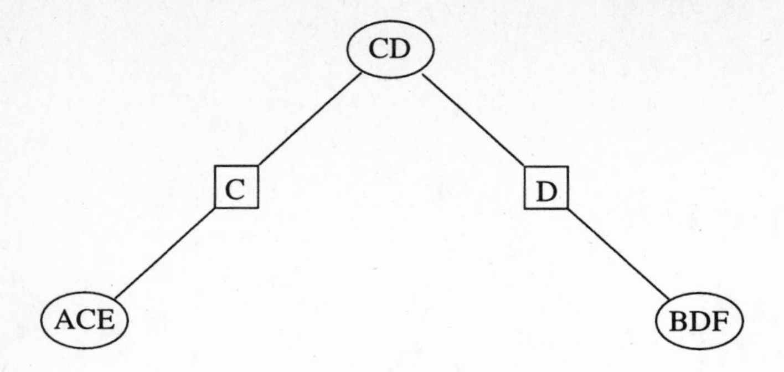

Figure 5.1 A consistent junction tree from which we want to calculate $P(A, B)$.

(ii) Calculate the joint probability for all possible transmitted words.

Stud farm. Go back to the example from Section 3.2.1 and the situation in Figure 3.9.

(i) Assume that Dorothy and Eric are about to mate. What is the chance that they are both AA?

(ii) The farm has to decide on a new mating among the horses Fred, Dorothy, Eric and Gwenn. Which pair should be chosen to minimize the risk of getting a carrier as an off-spring?

The first questions of both examples are rather easy to answer as they are requests for the probability of a specific configuration. This kind of question is answered by entering the configuration as evidence. By propagating, the requested probability is achieved through Theorem 4.8: take the sum of all entries in any table.

In this section we shall present methods for calculating probability tables for all configurations of a set of variables.

5.1.1 Two variables

In Figure 5.1 a situation is shown where we want to calculate $P(A, B)$. For convenience let us assume that the junction tree is consistent, that is

$$P(C) = \sum_{A,E} P(A, C, E) = \sum_{D} P(C, D)$$

and

$$P(D) = \sum_{C} P(C, D) = \sum_{B,F} P(B, D, F).$$

One way of calculating $P(A, B)$ we will call *firing of variables.* For each state a of A you enter a as a finding in (A, C, E) and call *CollectEvidence*(B, D, F). In this way $P(B, D, F, a)$ is achieved for all states a of A (Theorem 4.8), and therefore $P(B, D, F, A)$ as well as $P(A, B)$ are available.

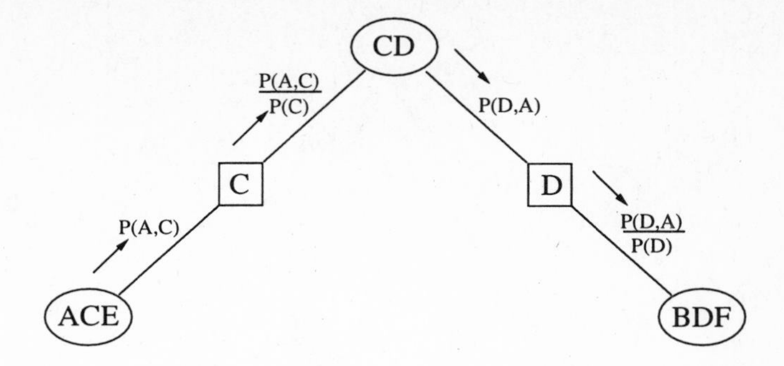

Figure 5.2 Propagation of variable A. The result is the table $P(A, B, D, F)$.

There is another method called *propagation of variables*. First $P(C, D, A)$ is calculated:

$$P(C, D, A) = P(D \mid A, C)P(A, C) = P(D \mid C)P(A, C) = P(C, D)\frac{P(A, C)}{P(C)}.$$

Next $P(A, B, D, F)$ is calculated:

$$\begin{aligned} P(A, B, D, F) &= P(B, F \mid D, A)P(D, A) \\ &= P(B, F \mid D)P(D, A) \\ &= P(B, D, F)\frac{P(D, A)}{P(D)}. \end{aligned}$$

The calculations above can be seen as a propagation as illustrated in Figure 5.2. The message $P(A, C)$ is passed from (A, C, F), the table of (C, D) is updated by the ratio $\frac{P(A,C)}{P(C)}$, and finally $P(D, A)$ is passed from (C, D) such that the table of (B, D, F) is updated by the ratio $\frac{P(D,A)}{P(D)}$.

5.1.2 An arbitrary set of variables

To illustrate the techniques, consider the junction tree in Figure 5.3, where the task is to calculate $P(A, B, C, D, E)$. For convenience we assume that the junction tree is consistent.

When firing variables, any node can be used as the collecting node. Usually a best node is one containing most of the space of the variable set. In this case the node (D, E, H) is the best choice. Now, any configuration (a, b, c) of states of A, B and C is entered as evidence in (A, J, K), (B, F) and (C, G) respectively, and *CollectEvidence* is called in (D, E, H). The result of each propagation is $P(D, E, H, a, b, c)$. By marginalization we get $P(D, E, a, b, c)$, and after 27 collections we have acquired $P(A, B, C, D, E)$.

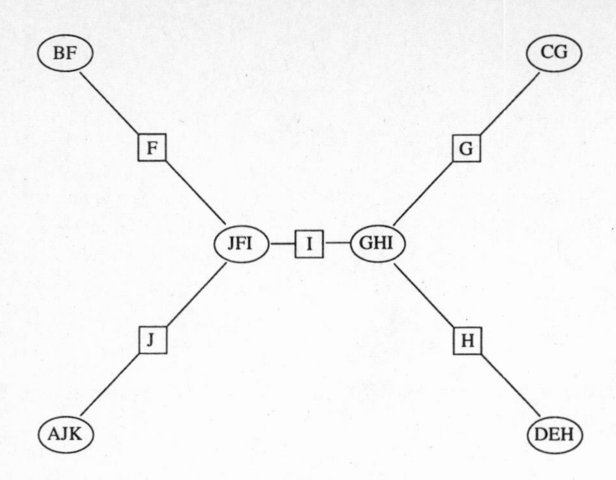

Figure 5.3 A junction tree from which we shall calculate $P(A, B, C, D, E)$. Each variable has three states.

Note that there is some redundance in this method. Let a be a state of A. For all configurations of (B, C) the same operations are performed when propagating $P(E, a)$ to (J, F, I). This redundance can be avoided by scheduling the set of propagations carefully: propagate $P(E, a)$ to (J, F, I) and keep this while the various states of B are propagated, etc.

Another way of avoiding the redundance is to perform propagation of variables. The calculations performed when propagating variables are:

- $P(A, E)$ is propagated to (J, F, I) by multiplying $P(J, F, I)$ by the ratio $\frac{P(A,E)}{P(E)}$. The result is $P(J, F, I, A)$.
- $P(B, F)$ is propagated to (J, F, I) by multiplying $P(J, F, I, A)$ by the ratio $\frac{P(B,F)}{P(F)}$. The result is $P(J, F, I, A, B)$.
- The table of (G, H, I) is updated by multiplying $P(G, H, I)$ by the ratios $\frac{P(A,B,I)}{P(I)}$ and $\frac{(C,G)}{P(G)}$. The result is $P(G, H, I, A, B, C)$.
- The table of $P(D, E, H)$ is updated by multiplying with the ratio $\frac{P(A,B,C,H)}{P(H)}$. The result is $P(D, E, H, A, B, C)$.

The proof of correctness is given in the next section.

5.1.3 Formalization of variable propagation

The heart of variable propagation is the following calibration operation.

Y-calibration. Let V and W be neighbours with separator S. Let $\mathbf{t}_W$ be a table over the variable set Z, and let Y be a subset of Z. V is said to *Y-calibrate to W*

through the following calculations.

$$\mathbf{t}_S^* = \sum_{Z \setminus (S \cup Y)} \mathbf{t}_W \qquad \mathbf{t}_V^* = \mathbf{t}_V \frac{\mathbf{t}_S^*}{\mathbf{t}_S}.$$

Note that $\mathbf{t}_S^*$ is a table over $S \cup Y$, and $\mathbf{t}_V^*$ becomes a table over $V \cup Y$.

Notice also that Vs table divided by Ss table is not changed by Y-calibration.

Variable propagation consists of a series of Y-calibrations from the leaves of a junction tree to a root. In Figure 5.3, for example, the node (D, E, H) is the root. The following proposition ensures correctness of variable propagation.

Proposition 5.1 *Let V and W be neighbours with separator S. Let Y be a set of variables such that $Y \cap V = \emptyset$. Suppose that $\mathbf{t}_V = P(V)$, $\mathbf{t}_W = P(Z)$ (where $Y \subseteq Z$) and $\mathbf{t}_S = P(S)$. Let V, Y-calibrate to W. Then $\mathbf{t}_V^* = P(V \cup Y)$.*

Proof. The result of the Y-calibration is

$$\mathbf{t}_V^* = P(V)\frac{P(S \cup Y)}{P(S)}.$$

Since $V \setminus S$ is independent of Y given S we have

$$P(V \setminus S \mid S \cup Y) = P(V \setminus S \mid S).$$

Therefore

$$\begin{aligned} P(V \cup Y) &= P(V \setminus S \mid S \cup Y)P(S \cup Y) \\ &= P(V \setminus S \mid S)P(S \cup Y) \\ &= \frac{P(V)}{P(S)}P(S \cup Y) \\ &= \mathbf{t}_V^*. \end{aligned}$$

□

***CollectEvidence*(V, X)**. Let V be a node in a junction tree representing $P(U)$ and let X be a set of variables. *CollectEvidence*(V, X) operates in the following way.

If *CollectEvidence*(V, X) is called from a neighbour V^*, *CollectEvidence*$(W, X \setminus V)$ is called in all neighbours W (except V^*). When *CollectEvidence*$(W, X \setminus V)$ has terminated with a table over Z, then V, Y-calibrates to W, where $Y = X \cap Z$.

The operation *CollectEvidence*(V, X) is used for calculating the joint probability table for X: V can be any node in the junction tree, and the result of *CollectEvidence*-(V, X) is that V holds $P(V \cup X)$.

In the example in Figure 5.3 the series of calls of *CollectEvidence* are:

CollectEvidence$((D, E, H), (A, B, C, D, E))$
CollectEvidence$((G, H, I), (A, B, C))$
CollectEvidence$((C, G), (A, B, C))$
CollectEvidence$((J, F, I), (A, B, C))$
CollectEvidence$((B, F), (A, B, C))$

Table 5.1 Probability tables for a small consistent system.

	a_1	a_2		b_1	b_2
b_1	0.3	0.2	c_1	0.1	0.35
b_2	0.1	0.4	c_2	0.4	0.15
	$P(A, B)$			$P(B, C)$	

$CollectEvidence((A, J, K), (A, B, C))$
(J, F, I) (A)-calibrates to (A, J, K)
(J, F, I) (B)-calibrates to (B, F)
(G, H, I) (A, B)-calibrates to (J, F, I)
(G, H, I) (C)-calibrates to (C, G)
(D, E, H) (A, B, C)-calibrates to (G, H, I).

From the proposition it is easy to see that $\mathbf{t}_{D,E,H} = P(A, B, C, D, E, H)$.

5.2 Configuration of maximal probability

In the example in Section 3.2.3 concerning transmission of symbol strings, the immediate task is to find out which symbol string most probably has been transmitted. Using propagation of variables the joint probabilities for all possible strings can be calculated and thereby the most probable string can be found. There is, however, a much more efficient method.

Example. Consider the small system consisting of the variables A, B and C with the joint probability determined by the consistent set of probabilities specified in Table 4.4, and suppose that we want to find out which configuration of (A, B, C) has maximal probability.

We have that the (A, B)-configuration of maximal probability is (a_2, b_2), and the (B, C)-configuration of maximal probability is (c_2, b_1). They cannot be combined into one configuration, and it seems that we have to calculate $P(A, B, C)$ and determine the maximum of that table. Let us look at the calculation

$$\begin{aligned}
\max_{A,B,C} P(A, B, C) &= \max_{A,B,C} \frac{P(A, B)P(B, C)}{P(B)} \\
&= \max_{A,B} \left(\max_{C} \frac{P(A, B)P(B, C)}{P(B)} \right) \\
&= \max_{A,B} \frac{P(A, B)}{P(B)} \max_{C} P(B, C) \\
&= \max_{A,B} P(A, B) \frac{\max_C P(B, C)}{P(B)} \\
&= \max_{A,B} (P(A, B)(0.8, 0.7)) \\
&= \max_{A,B}(0.24, 0.16, 0.07, 0.28) \\
&= 0.28.
\end{aligned}$$

So, the most probable configuration has the probability 0.28, and it has the states (a_2, b_2) at the first two places. It is now easy to check in the table $P(b_2, C)$ that C must be in state c_1.

Note that in the calculation above the joint table $P(A, B, C)$ was never calculated.

We shall now formalize the calculations as a propagation method for finding a most probable configuration (abbreviated MPC). Before we do so, we shall define precisely the concept of configuration.

Definition. Let V be a set of variables. A *configuration* v^* of V is a set of states $\{a, \ldots, b\}$ which contains exactly one state from each variable of V.

Let W be a subset of V, and let w^* be a configuration of W. Then $V * w^*$ is the set of configurations of V containing w^*:

$$V * w^* = \{v^* \mid v^* \text{ a configuration of } V, w^* \subseteq v^*\}.$$

The propagation method is based on the following calibration operation.

Max-calibration

Let V and W be neighbours with separator S, and with tables $\mathbf{t}_V$, $\mathbf{t}_W$ and $\mathbf{t}_S$ respectively. V is said to *Max-calibrate* to W through the following calculations:

$$\mathbf{t}_S^* = \max_{W \setminus S} \mathbf{t}_W \qquad \mathbf{t}_V^* = \mathbf{t}_V \frac{\mathbf{t}_S^*}{\mathbf{t}_S}.$$

So, the only difference between normal calibration and Max- calibration is that $\sum$ has been changed to max.

Note that the calculations in the example contains a Max-calibration. Note also that the product of the two clique tables divided by the separator table is invariant under Max-calibration.

Max-consistence

Two neighbours V and W with separator S in a junction tree are *Max-consistent* if

$$\max_{V \setminus S} \mathbf{t}_V = \mathbf{t}_S = \max_{W \setminus S} \mathbf{t}_W.$$

Note. If first V Max-calibrates to W, and afterwards W Max-calibrates to V, then they are Max-consistent.

Max-marginal

Let V be a set of variables, and let $W \subseteq V$. $\mathbf{t}_W$ is the *Max-marginal* of $\mathbf{t}_V$ if

$$\mathbf{t}_W = \max_{V \setminus W} \mathbf{t}_V.$$

That is, $\mathbf{t}_W(w^*) = \max_{y^* \in V * w^*} \mathbf{t}_V(y^*)$.

Max-propagation

Max-propagation is like HUGIN propagation, except that Max-calibration is used instead of normal calibration. So, Max-propagation consists of a *MaxCollect* followed by a *MaxDistribute* from the same clique.

The result of Max-propagation is that the tables hold the appropriate Max-marginals of the full table. The proof is very similar to the proof of correctness of HUGIN propagation. It is sketched below.

Theorem 5.1 *The product of all clique tables divided by all separator tables is invariant under Max-propagation.*

Proof. See the note after the definition of Max-calibration. □

Theorem 5.2 *Max-propagation results in a Max-consistent junction tree.*

Proof. As the proof of Theorem 4.4(ii). Use that $\max_Y \mathbf{t}_X\mathbf{t}_Y = \mathbf{t}_X \max_Y \mathbf{t}_Y$ whenever $X \cap Y = \emptyset$. □

Theorem 5.3 *Let T be a Max-consistent junction tree over the universe U. Let $\mathbf{t}_U$ be the product of all clique tables divided by all separator tables. Let V be a node with table $\mathbf{t}_V$, and let S be a separator with table $\mathbf{t}_S$. Then*

$$\mathbf{t}_V = \max_{U \setminus V} \mathbf{t}_U \qquad\qquad \mathbf{t}_S = \max_{U \setminus S} \mathbf{t}_U .$$

Proof. Similar to the proof of Theorem 4.7. □

Corollary 5.1 *Let T be a junction tree over U with $\mathbf{t}_U = P(U, e)$, where e is evidence entered (but not necessarily propagated). Perform a Max-propagation in T.*

(i) Then for any clique V we have $\mathbf{t}_V = \max_{U \setminus V} P(U, e)$.

(ii) For any configuration x^* of V we have $\mathbf{t}_V(x^*) = \max_{y^* \in U^* x^*} P(y^*, e)$. That is, $\mathbf{t}_V(x^*)$ is the maximum of $P(u^*, e)$ over all configurations with $x^* \subseteq u^*$.

Furthermore, let v^* be a configuration of V such that $\mathbf{t}_V(v^*) = \max_V \mathbf{t}_V$. Then

(iii) $\mathbf{t}_V(v^*) = \max_U P(U, e)$.

(iv) There is a configuration u^* of U such that $\mathbf{t}_U(u^*) = \max_U P(U, e)$ and $v^* \subseteq u^*$.

The above also holds for separators.

Proof. Since $\mathbf{t}_U = P(U, e)$, then *(i)* is a result of Theorem 5.3.
(ii) is a result of *(i)* and the definition of $U * x^*$.
(iii) $\mathbf{t}_V(v^*) = \max_V \mathbf{t}_V = \max_V(\max_{U \setminus V} P(U, e)) = \max_U P(U, e)$.
(iv) From *(ii)* and *(iii)*. □

Corollary 5.1 gives a method for determining configurations of maximal probability. First perform a Max-propagation. If each table in the junction tree has exactly one configuration of highest value then it is part of the single MPC. If some tables have several configurations of maximal value, then there are also several MPCs. To determine one MPC you can enter one of the configurations of maximal value as evidence and perform a new Max-propagation. Repeat this until the result is a single MPC.

Note that you can always change between Max-propagation and HUGIN propagation. By performing a Max-propagation the product of all clique tables divided by all separator tables remains equal to $P(U, e)$. So, if you perform a HUGIN propagation afterwards, Theorem 4.8 ensures that the tables in the cliques and separators are the appropriate marginals of $P(U, e)$.

5.3 Data conflict

A Bayesian network represents a closed world with a finite set of variables and causal relations. The causal relations are not universal, but reflect relations under certain constraints. Take, for example, a diagnostic system which on the basis of blood analysis monitors pregnancy. Only diseases and relations relevant for pregnant women are represented in the model. So, if the blood originates from a man, the case is not covered by the model. It may happen that findings from male blood are impossible given the model. If so, the inconsistence is easy to detect: the probability of the evidence is zero. However, more often a set of findings is possible in the given model, and the system will not object to it. It will yield posterior probability distributions which may look rather harmless. The same also happens if test results are flawed. In a diagnostic situation a single flawed test result may take the investigation in a completely wrong direction (such flawed pieces of information are called red herrings).

5.3.1 Seismometer

Watson makes frequent calls to Holmes regarding the burglar alarm, however, till now the cause of activation of the alarm has been small earthquakes (cf. the example in Section 2.1.4). Every time Holmes rushes home, just to find that everything is in order; so now Holmes is installing a seismometer in his house with a direct line to his office. The seismomenter has three states:

0 for no vibrations;
1 for small vibrations (caused by minor earthquakes or passing trucks);
2 for larger vibrations (caused by major earthquakes or persons walking around in his house).

The network for this alarm system is shown in Figure 5.4, and the tables are given in Table 5.2.

One afternoon Watson calls again and announces that Holmes' alarm has gone off. Holmes checks the seismometer, it is in state 0! From our knowledge of the network model we would say that the findings are in conflict. However, an evidence propagation does not disclose that. The posterior probability for burglary is 0.48.

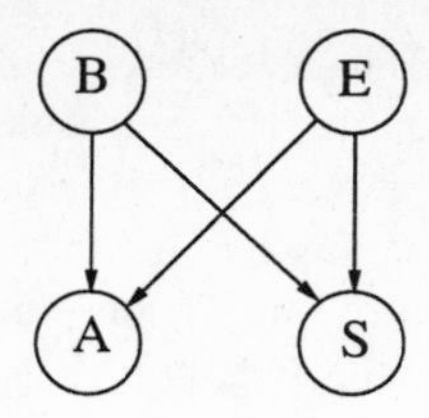

Figure 5.4 Mr Holmes' alarm system with seismometer (A for alarm, B for burglary, E for earthquake, and S for seismometer).

Table 5.2 Conditional probabilities for the seismometer model. Prior for B: (0.5, 0.5), and for E: (0.1, 0.9).

	$E = y$	$E = n$
$B = y$	(0, 0.03, 0.97)	(0.01, 0.02, 0.97)
$B = n$	(0.01, 0.97, 0.02)	(0.97, 0.02, 0.01)

$P(S \mid E, B)$

	$E = y$	$E = n$
$B = y$	(1, 0)	(0.99, 0.01)
$B = n$	(0.99, 0.01)	(0.01, 0.99)

$P(A \mid E, B)$

The problem for Holmes is whether Watson is playing a practical joke, whether the seismometer is defective, or whether the problem is that he is dealing with a very rare case.

Using the model only, you cannot distinguish between a case not covered by the model and flawed data. So, what we can do is to provide Holmes with a measure that indicates possible conflicts between data and model.

5.3.2 The conflict measure (conf)

Several approaches for analyzing data for conflicts are made. In this section we shall present a measure which is easy to calculate and which gives an indication of a possible conflict. The idea behind the measure is that correct findings originating from a coherent case covered by the model conform to certain expected patterns laid down in the model. In other words, the findings should be positively correlated. If $e = \{x, \ldots, y\}$ is a set of findings, we would expect $P(e)$ to exceed the probability for independent findings: $P(x) \cdots P(y)$. Hence we define the *conflict measure* as

$$\text{conf}(\{x, \ldots, y\}) = \log \frac{P(x) \cdots P(y)}{P(e)}.$$

The reason for the log is sheer convenience; some formulas look nicer. You can take any logarithm. Here the log with base 2 is chosen.

A positive conf(e) is an indicator of a possible conflict. For the seismometer case the conf-value is 4.7.

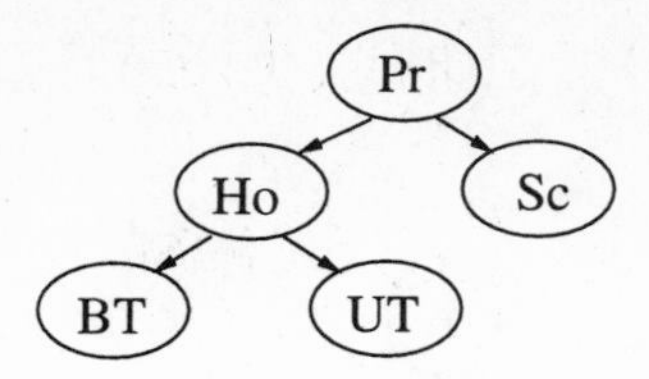

Figure 5.5 A diagnostic model for pregnancy.

The probabilities for the individual findings in the formula for conf can be read directly from the initial network, and the joint probability of the entire set of findings can be calculated through Theorem 4.8.

5.3.3 Tracing of conflicts

After the conflict measure has been found to be positive, a further task would then be to find out whether a possible conflict is due to flawed findings – and if so, to trace them.

To illustrate this problem we shall use the insemination example from Section 3.1.3. The model is given in Figure 5.5, and the tables can be seen in Exercise 3.1.

Suppose we have the evidence $e = \{BT = y, UT = y, Sc = n\}$. Seen from outside the set seems to be conflicting. $P(e)$ is 0.0447. Initially $P(BT = y) = 0.57$, $P(UT = y) = 0.63$, $P(Sc = n) = 0.22$, and so

$$\text{conf}(e) = 0.82.$$

Though the conflict is not very large, we may want to trace the origin of it further. The evidence e is communicated to Pr in two sets, $e' = \{BT = y, UT = y\}$ and $e'' = \{Sc = n\}$. A further investigation could therefore be to see whether e' in itself is in conflict. To do that we need $P(e')$, which is 0.4394. We get

$$\text{conf}(e') = -0.29,$$

and not surprisingly no conflict is detected in e'.

Another possibility could be that the two sets e' and e'' are conflicting. We define

$$\text{conf}(e', e'') = \log \frac{P(e')P(e'')}{P(e)} = 1.11,$$

which indicates that the two sets of findings are conflicting, and we conclude that either e' or e'' is flawed (most probably e'').

We shall now deal with tracing of conflicts in more general terms. Figure 5.6 shows a junction tree with findings $e = \{x, y, z, u, v\}$ entered. If *CollectEvidence*(V) is called, then for any node in the junction tree (except for V) some evidence e' is collected and transmitted further up. The conflict conf(e') is called a *partial conflict*. Also the evidence collected in a node W consists of sets of findings transmitted along

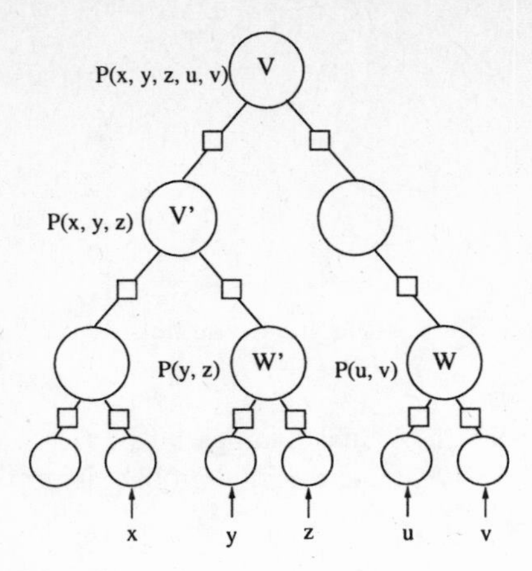

Figure 5.6 A junction tree with findings x, y, z, u, and v entered. Theorem 4.8 provides the joint probabilities indicated at the nodes V, V', W, and W'.

the links to W. If these sets are $e_1, \ldots, e_n$ we define the *local conflict* as

$$\text{conf}(\{e_1, \ldots, e_n\}) = \log \frac{P(e_1) \cdots P(e_n)}{P(e_1 \cdots e_m)}.$$

To calculate partial as well as local conflicts we need $P(e')$ where e' is the evidence collected at a node W. At the end of Section 4.3.1 it was shown how $P(e')$ is accessed. Figure 5.6 indicates the probabilities directly obtainable through a call of *CollectEvidence*(V).

To round off this section we give the following proposition which relates the three kinds of conflicts.

Proposition 5.2 *Let* CollectEvidence(V) *result in collection of the evidence e arriving to V as* $\{e_1, \ldots, e_n\}$. *Then*

$$\text{conf}(e) = \sum_i \text{conf}(e_i) + \text{conf}(\{e_1, \ldots, e_n\}).$$

In words: the global conflict is the sum of the local conflict and the corresponding partial conflicts.

Proof. Exercise 5.8. □

5.3.4 Conflict or rare case

It may happen that typical data from a very rare case causes a high conf-value. In the case of Holmes' alarm system we could add *Flood* as a cause for the alarm to go off, and *Flood* can be caused by *Rain* (see Fig. 5.7).

For this system we get $\text{conf}(\{A = y, S = 0\}) = 4.6$. It is still indicating a possible conflict. The reason is that though the evidence is perfectly coherent in the case

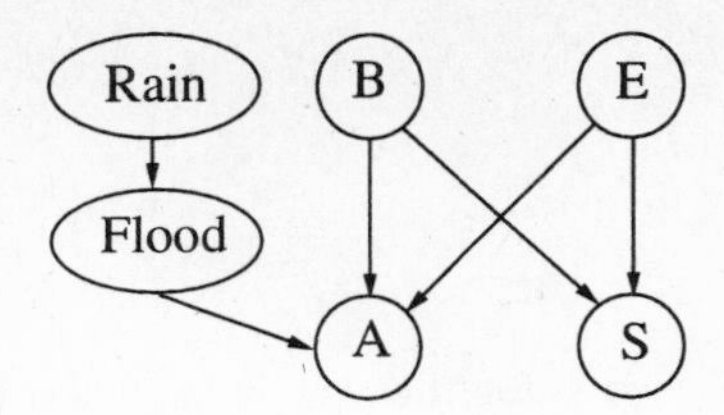

Figure 5.7 Holmes' revised network. $P(Rain) = (0.01, 0.99)$; $P(Flood = y \mid Rain = y) = 0.1$; $P(Flood = y \mid Rain = n) = 0$. $P(A \mid \text{parents})$ are $(1, 0)$ if two parents are y, it is $(0.99, 0.01)$ if one parent is y, and it is $(0.01, 0.99)$ if all parents are n.

of flood, the occurrence of flood is very rare. Holmes looks out of his window. It rains cats and dogs, and he has resolved his problem; the model gives the posterior probability for flood, 0.84.

The problem above calls for a method for pointing out whether a positive conf-value may be explained as a rare case covered by the model.

Let $e = \{x, \dots, y\}$ be findings for which $\text{conf}(e) > 0$, and let h be a hypothesis which could explain the findings: $\text{conf}(\{x, \dots, y, h\}) \leq 0$.

We have

$$\text{conf}(\{x, \dots, y, h\}) = \log \frac{P(x) \cdots P(y) P(h)}{P(e, h)} = \text{conf}(e) + \log \frac{P(h)}{P(h \mid e)}.$$

This means that if

$$\log \frac{P(h \mid e)}{P(h)} \geq \text{conf}(e) \tag{5.1}$$

then h can explain away the conflict. In the flood example the value of the lefthand side of (5.1) is 5.6.

The fraction $\frac{P(h|e)}{P(h)}$ is used in various ways, and it is called the *normalized likelihood*. Note that by Bayes' rule

$$\frac{P(h \mid e)}{P(h)} = \frac{P(e \mid h)}{P(e)}.$$

Normalized likelihoods can be monitored automatically for all variables. Therefore, when analyzing for conflict/rare case, it is easy to detect whether a conflict may be due to a particular variable being in a very rare state.

5.3.5 Other approaches to conflict detection

The conf measure is not the only way of dealing with conflict detection. An approach to the problem would be to incorporate sources of surprise directly into the model. This can be done by entering variables modelling probabilities for malfunctioning of sensors, and to extend causal variables like disease variables with the state *other*. This approach, however, has the problem that it is difficult to model malfunctions

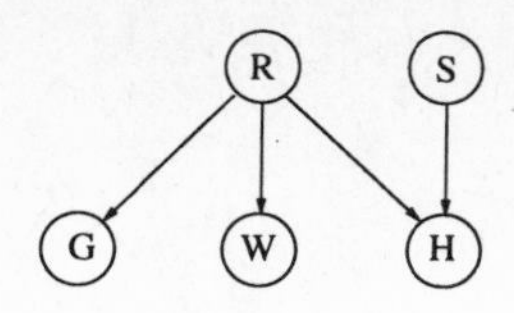

Figure 5.8 Network for the wet grass example. Holmes can inspect both Watson's and Mrs Gibbon's grass.

or *other* unless the types of malfunction and *other* are known. Also, with *other* you can only handle discrepancies which are local in the network.

Another approach is to calculate a so-called *surprise index* for the set of findings. If the findings e are statements of the variables $A, \ldots, B$, the surprise index is the sum of probabilities for all configurations of $(A, \ldots, B)$ with a probability no higher than $P(e)$. If the surprise index is less than 0.1, this should be an indication of a possible conflict. In the seismometer case, the surprise index is 0.03. Unfortunately, the calculation of a surprise index is exponential in the number of findings, and it must be considered intractable in general.

5.4 Sensitivity analysis

Evidence e has been entered to a network, and some hypotheses $h_1, \ldots, h_n$ are in the focus of interest. Sensitivity analysis shall give answers to questions like:
- which evidence is in favour of/against/irrelevant for h_i?
- which evidence discriminates h_i from h_j?

5.4.1 Example and definitions

The following variant of the wet grass example is used for illustration.

> In the morning when Mr Holmes leaves his house he realizes that his grass is wet. He wonders whether it has rained during the night or whether he has forgotten to turn off his sprinkler. He looks at the grass of his neighbours, Dr Watson and Mrs Gibbon. Both lawns are dry and he concludes that he must have forgotten to turn off his sprinkler.

The network for Holmes' reasoning is shown in Figure 5.8, and the initial probabilities are given in Table 5.3.

The evidence e consists of the three findings e_H, e_W, e_J, and the hypothesis in focus is h_s : "$S = y$". We have $P(h_s) = 0.1$, and $P(h_s \mid e) = 0.9999$.

We have $P(h_s \mid e_H) = 0.51$, $P(h_s \mid e_W) = 0.1 = P(h_s \mid e_J)$. (A d-separation analysis could yield some of the results, however, this is not the point here.) So neither e_W nor e_J alone have any impact on the hypothesis, but e_H is not sufficient for the conclusion. Therefore the immediate conclusion that e_W and e_J are irrelevant for the hypothesis is not correct. We must conclude that evidence in combination may have a larger impact than the "sum" of the individual impacts.

Table 5.3 Tables for the wet grass example. $P(R) = (0.1, 0.9) = P(S)$.

	$R = y$	$R = n$		$R = y$	$R = n$
y	0.99	0.1	$S = y$	(1, 0)	(0.9, 0.1)
n	0.01	0.9	$S = n$	(0.99, 0.01)	(0, 1)
$P(G \mid R) = P(W \mid R)$				$P(H \mid R, S)$	

Table 5.4 Normalized likelihoods for the subsets in the example. A "1" in the table indicates that the finding is an element of e'.

$W = n$	$J = n$	$H = y$	$\frac{P(e' \mid h_s)}{P(e)}$
1	1	1	9.999
1	1	0	1
1	0	1	9.88
1	0	0	1
0	1	1	9.88
0	1	0	1
0	0	1	5.1
0	0	0	1

To investigate further we must consider the impacts of subsets of the evidence. We have

$$P(h_s \mid e_W, e_J) = 0.1 \quad P(h_s \mid e_H, e_J) = 0.988 = P(h_s \mid e_W, e_H).$$

To relate the probabilities above to their impact on the hypothesis we can divide them with the prior probability $P(h_s)$ to get the normalized likelihood.

Other measures can be used, for example *Bayes' factors*:

$$\frac{P(e \mid h)}{P(e \mid \neg h)}$$

or the fraction of achieved probability

$$\frac{P(h \mid e')}{P(h \mid e)}.$$

The various normalized likelihoods are given in Table 5.4.

From Table 5.4 we can conclude that no single finding is sufficient for the conclusion. Also, though (e_W, e_J) alone has no impact on h_s, these two findings cannot both be removed. Moreover, we see that the subsets (e_H, e_J) and (e_H, e_W) can account for almost all the change in the probability for h_s.

Definitions. Let e be evidence and h a hypothesis. Suppose that we want to investigate how sensitive the result $P(h \mid e)$ is to the particular set e.

We shall say that evidence $e' \subseteq e$ is *sufficient* if $P(h \mid e')$ is almost equal to $P(h \mid e')$. We then also say that $e \setminus e'$ is *redundant*. The term *almost equal* can be made precise by selecting a threshold θ_1 and require that $\left|\frac{P(h|e')}{P(h|e)} - 1\right| < \theta_1$. Note that $\frac{P(h|e')}{P(h|e)}$ is the fraction between the two likelihood ratios.

e' is *minimal sufficient* if it is sufficient, but no proper subset of e' is so.

e' is *crucial* if it is a subset of all sufficient sets.

e' is *important* if the probability of h changes too much without it. To be more precise: if $\left|\frac{P(h|e \setminus e')}{P(h|e)} - 1\right| > \theta_2$, where θ_2 is some chosen threshold.

In the example above put $\theta_2 = 0.2, \theta_1 = 0.05$. Then (e_H, e_J) and (e_H, e_W) are minimal sufficient, (e_W, e_J) is important, and e_H is crucial.

In Holmes' universe there is another possible hypothesis, namely h_r : "$R = y$". To find out which findings discriminate between the two hypotheses an analysis of r can be performed. $P(h_r \mid e')$ is calculated for each subset of e', and the ratio between the two (normalized) likelihoods is used. The ratios are shown in Table 5.5.

Table 5.5 Likelihood ratios for the hypotheses h_s and h_r.

$W = n$	$J = n$	$H = y$	$\frac{P(e' \mid h_s)}{P(e' \mid h_r)}$
1	1	1	6622
1	1	0	7300
1	0	1	74
1	0	0	81
0	1	1	74
0	1	0	81
0	0	1	0.92
0	0	0	1

Table 5.5 shows that e_W and e_J are good discriminators between the two hypotheses.

As illustrated above, the heart of sensitivity analysis is the calculation of $P(h \mid e')$ for each $e' \subseteq e$. Since the number of subsets grows exponentially with the number of findings, the job may become very heavy, particularly when $P(h \mid e')$ has to be calculated through a propagation in a large network. Therefore our task is to *construct methods for the calculation of* $P(h \mid e')$ *for as many subsets as possible without running into an exponential job.* We shall call a subset e' *accessed* if $P(h \mid e')$ is achieved.

From the considerations at the end of Section 4.4.1 we get access to some $P(e')$s. To get $P(e' \mid h)$ for the same subsets we simply enter and propagate h before propagating e. Then we get $P(h)$ and all tables are conditioned by h.

Finally, Bayes' rule yields

$$P(h \mid e') = \frac{P(e' \mid h)P(h)}{P(e')}.$$

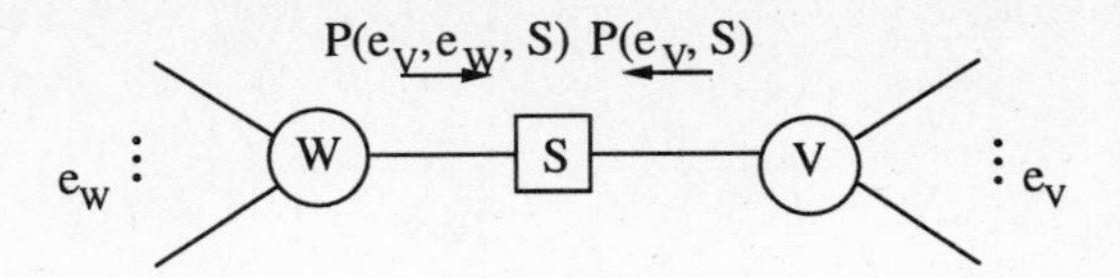

Figure 5.9 The transmitted tables in a network where *CollectEvidence* is called at the left of W.

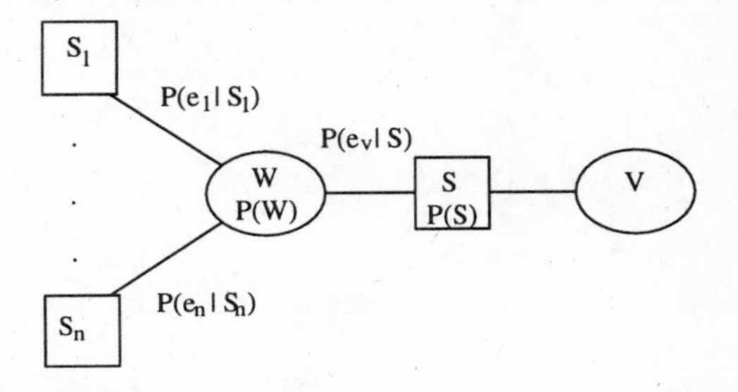

Figure 5.10 The situation when W is about to pass $P(e_W, S)$ to S.

Note that when $P(h \mid e')$ and $P(h)$ are available then Bayes' factors can also be calculated:

$$\frac{P(e' \mid h)}{P(e' \mid \neg h)} = \frac{P(h \mid e')P(\neg h)}{P(h)P(\neg h \mid e')} = \frac{P(h \mid e')(1 - P(h))}{P(h)(1 - P(h \mid e'))}.$$

5.4.2 Cautious propagation

Consider Figure 5.9 and observe that in the *DistributeEvidence* phase, $P(V, e_V)$ is multiplied by $\frac{P(e_V, e_W, S)}{P(e_V, S)}$. If instead we multiply by $P(e_W \mid S)$ the only difference will be multiplications for configurations s, where $P(s, e_V) = 0$. In that case the corresponding entries in $P(V, e_V)$ are also zero. So multiplying by $P(e_W \mid S)$ will give exactly the same result as given by HUGIN propagation.

Therefore, the idea behind cautious propagation is that in the *CollectEvidence* phase let S keep $P(S)$, and store $P(W)$ and $P(e_V \mid S)$ separately near W. Now, assume that W has received and stored the messages $P(e_1 \mid S_1), \ldots, P(e_n \mid S_n)$ (see Fig. 5.10), where $e_W = e_1 \cup \cdots \cup e_n$.

From HUGIN propagation we know that

$$P(W, e_W) = P(W)\frac{P(S_1, e_1)}{P(S_1)} \cdots \frac{P(S_n, e_n)}{P(S_n)},$$

and therefore

$$P(e_W, S) = \sum_{W \setminus S} P(W)P(e_1 \mid S_1) \cdots P(e_n \mid S_n),$$

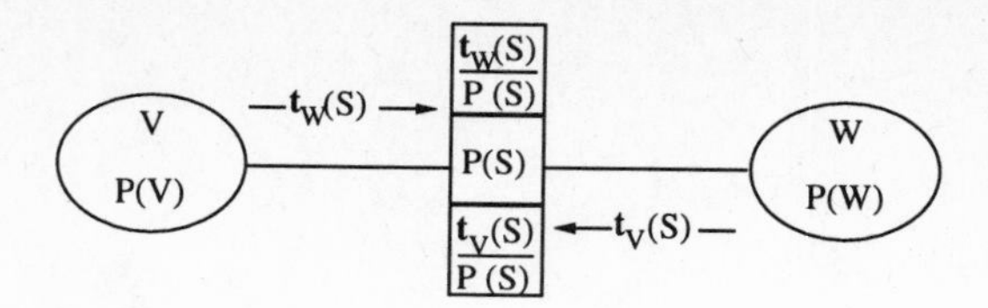

Figure 5.11 A link in a junction tree and the tables stored when a cautious message has been passed in both directions.

and

$$P(e_W \mid S) = \frac{P(e_W, S)}{P(S)}.$$

If W is the root, Figure 5.10 describes the situation after *CollectEvidence*(W), and recursively it will, in the *DistributeEvidence* phase, be the situation for all non-leaf nodes.

A more precise description of cautious propagation is the following.

Before evidence is entered, each separator S holds $P(S)$ and each clique V holds $P(V)$. Whenever S receives a table $\mathbf{t}_V(S)$ from a neighbour W, it is divided by $P(S)$, stored, and a message is sent to the other neighbour clique V (see Fig. 5.11).

A clique W can send a table to a neighbour separator S if it has received a message from all its other neighbouring separators $S_1, \ldots, S_n$. The table sent is

$$\mathbf{t}_V(S) = \sum_{W \setminus S} P(W) \frac{\mathbf{t}_W(S_1)}{P(S_1)} \cdots \frac{\mathbf{t}_W(S_n)}{P(S_n)}.$$

For the correctness of the method we see from the considerations above that it is enough to show that $\mathbf{t}_V(S) = P(S, e_W)$, where e_W is the evidence entered at the subtree containing S and W, but not V.

If W is a leaf then

$$\mathbf{t}_V(S) = \sum_{W \setminus S} P(W, e_W) = P(S, e_W).$$

If $\mathbf{t}_W(S_i) = P(S_i, e_i)$ then $\frac{\mathbf{t}_W(S_i)}{P(S_i)} = P(e_i \mid S_i)$, and because $e_1, \ldots, e_n$ are independent given W we have (see Fig. 5.10)

$$\begin{aligned} P(e_W \mid W) &= P(e_1 \mid W) \cdots P(e_n \mid W) \\ &= P(e_1 \mid S_1) \cdots P(e_n \mid S_n). \end{aligned}$$

Therefore

$$\begin{aligned} \mathbf{t}_V(S) &= \sum_{W \setminus S} P(W) \frac{\mathbf{t}_W(S_1)}{P(S_1)} \cdots \frac{\mathbf{t}_W(S_n)}{P(S_n)} \\ &= \sum_{W \setminus S} P(W) P(e_i \mid S_1) \cdots P(e_n \mid S_n) \end{aligned}$$

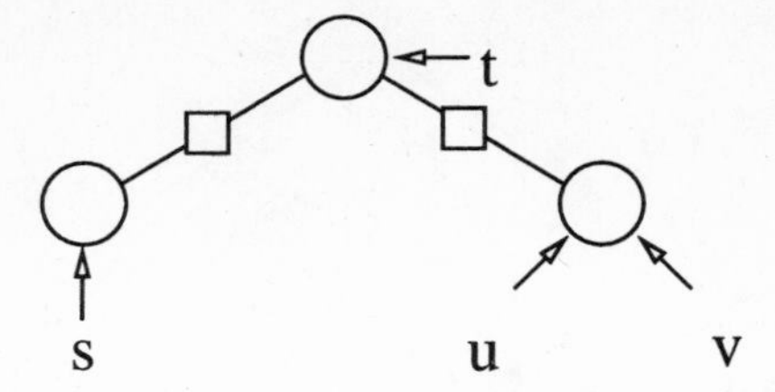

Figure 5.12 A set of findings, where $P(s, u, v)$, $P(s, t, u)$ and $P(s, t, v)$ cannot be accessed directly.

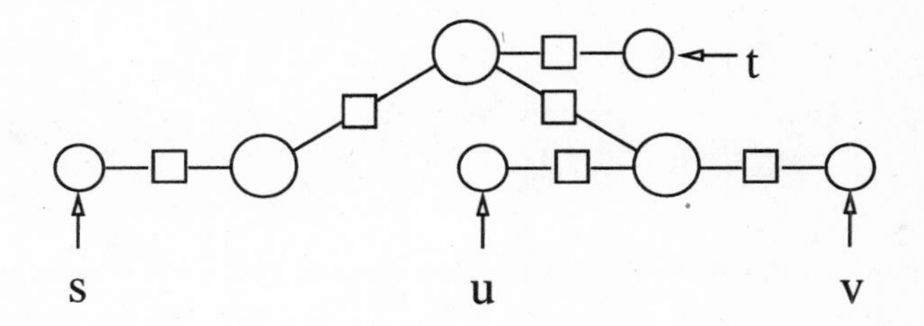

Figure 5.13 Now the probability of all complements of single findings can be accessed by cautious propagation.

$$= \sum_{W \setminus S} P(W)P(e_W \mid W)$$

$$= \sum_{W \setminus S} P(W, e_W)$$

$$= P(S, e_W).$$

Note. Any separator divides the evidence into two sets e_1 and e_2. Cautious propagation has the side effect that both $P(e_1)$ and $P(e_2)$ are easy to calculate.

Cautious entering of evidence

Consider the situation in Figure 5.12. We can access the complement of s. However, with the present techniques we cannot access the complements of $\{t\}$, $\{u\}$ and $\{v\}$. Fortunately there is an easy way out: add dummy variables such that findings are always inserted to a leaf in the junction tree, and such that at most one finding is inserted in any node. The construction is illustrated in Figure 5.13.

This structural way of accessing complements of single findings through propagation can be achieved in a more simple manner by the following modification of the way that findings are inserted. Instead of changing the table in the clique when inserting the finding f, consider the finding as a message $\underline{f}$, store it, and treat it in the same way as tables from neighbour separators.

In the following we include cautious entering of evidence in the term *cautious propagation*.

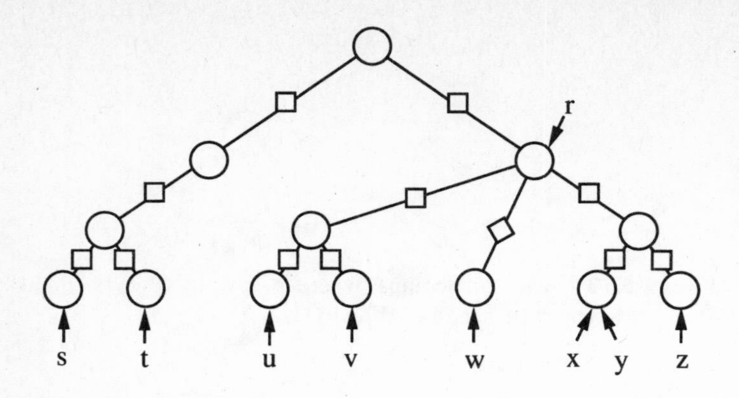

Figure 5.14 A junction tree with evidence $e = \{r, s, t, u, v, w, x, y, z\}$ entered. The top node is the chosen root.

$P(e')$s provided by cautious propagation

For the situation in Figure 5.14 we can directly access the probability of the following sets:
$\{s\}, e \setminus \{s\}, \{t\}, e \setminus \{t\}, \{u\}, e \setminus \{u\}, \{x\}, \{y\}, \{r\}, \{v\}, e \setminus \{v\}, \{w\}, e \setminus \{w\}, \{x, y\}, e \setminus \{x, y\}, \{z\}, e \setminus \{z\}, \{s, t\}, e \setminus \{s, t\}, \{u, v\}, e \setminus \{u, v\}, \{r, s, t, u, v, w\}, \{x, y, z\}$.

Let V be a clique with findings $f_1, \ldots, f_m$ entered and with adjacent separators $S_1, \ldots, S_n$. Let e_i denote the evidence entered to the subtree containing S_i but not V. Take, for example, the set $e_1 \cup e_2 \cup \{f_1\}$. Then

$$P(V, e_1, e_2, f_1) = P(V)P(e_1 \mid S_1)P(e_2 \mid S_2)\underline{f}_1$$

and all the factors in the product are available local to V. Therefore $P(e_1 \cup e_2 \cup \{f_1\})$ is easy to calculate.

In general, cautious propagation gives access to the probability of any union of the sets $\{f_1\}, \ldots, \{f_m\}, e_1, \ldots, e_n$.

In Figure 5.14 we therefore also get access to the probability of
$e\setminus\{x\}, e\setminus\{y\}, e\setminus\{r\}, \{s, t, r\}, \{u, v, w, x, y, z\}, \{s, t, u, v\}, \{r, w, x, y, z\}, \{s, t, w\}, \{r, u, v, x, y, z\}, \{s, t, x, y, z\}, \{r, u, v, w\}, \{r, s, t, u, v\}, \{w, x, y, z\}, \{r, s, t, w\}, \{u, v, x, y, z\}, \{r, s, t, x, y, z\}, \{u, v, w\}, \{s, t, u, v, w\}, \{r, x, y, z\}, \{s, t, u, v, x, y, z\}, \{r, w\}, \{s, t, w, x, y, z, \}, \{r, u, v\}$.

5.4.3 "What if"-questions, the set of crucial findings and minimal sufficient sets of findings

Consider the situation where evidence e has been entered and propagated cautiously. Assume that h is in the focus of interest, and that e has also been propagated in the junction tree conditioned on h.

Now, suppose we want to investigate the impact on h if a finding x is changed to y. This can be done for any finding without a new propagation. Let V be the clique to which the finding is entered. Due to cautious propagation, the situation is so that local to V we have $P(V)$, $P(e_i \mid S_i)$ for all adjacent separators S_i, and tables f for the findings f to multiply on $P(V)$. It is then easy to substitute x with y calculate $P(V, e \cup \{y\} \setminus \{x\})$ as the product of all tables local to V, and finally to marginalize V out to get $P(e \cup \{y\} \setminus \{x\})$. The same is done with the junction tree conditioned on h to get $P(e \cup \{y\} \setminus \{x\} \mid h)$. Bayes' rule now yields $P(h \mid \mathrm{e} \cup \{y\} \setminus h)$.

Next, assume that the posterior probability of h is high and we want to determine the set of crucial findings. It may happen that some findings are evidence against h, but that they are overwritten by the entire set. We assume that the findings acting against h have been sorted out. For the remaining evidence we then assume *monotonicity*: no non-sufficient set contains a sufficient subset.

If no proper subsets of e are sufficient then all findings are crucial, else the set of crucial findings is the intersection of all sufficient sets of the form $e \setminus \{x\}$. Suppose that x is not crucial, then due to monotonicity $e \setminus \{x\}$ is sufficient. On the other hand, if x is crucial then $e \setminus \{x\}$ cannot be sufficient.

When e is propagated cautiously, $P(h \mid e \setminus \{f\})$ can be calculated for all f, and thereby the set of crucial findings can be determined as the intersection of the sufficient subsets accessed.

A procedure for finding a minimal sufficient set of findings could be the following depth-first search.

(1) Among the subsets e' accessed through cautious propagation, choose a sufficient set e_1^* with a minimal number of findings.

(2) Enter e_1^* and perform a cautious propagation in an initialized junction tree and in the one conditioned on h.

(3) Perform steps (1) and (2) until no sufficient subset is accessed. The set entered last is a minimal sufficient set of findings.

The procedure terminates correctly due to the monotonicity assumption and the fact that all complements of single findings are accessed.

Take as an example the situation in Figure 5.14, and assume that $\{s, w, z\}$ is minimal sufficient. The search will then first identify $\{s, t, w, x, y, z\}$ as sufficient, next $\{s, w, z\}$ is identified, and finally it is concluded that $\{s, w, z\}$ is minimal sufficient. The search requires three pairs of cautious propagation.

The worst case for this search occurs when the evidence is entered as pairs of findings in leafs, and a minimal sufficient set consists of exactly one finding from each pair (see Fig. 5.15).

In that case, only one finding is removed in each round, and the search requires $n/2 + 1$ pairs of propagations.

If we want all minimal sufficient subsets, a breadth-first search can be performed. If, in the example of Figure 5.14, $\{s, w, z\}$, $\{t, w, z\}$, $\{s, w, x, y\}$ and $\{t, w, x, y\}$ are minimal sufficient, then the sets to propagate are the following:

- initially e;
- first round: $\{s, t, w, x, y, z\}$, $\{r, t, u, v, w, x, y, z\}$, $\{r, s, u, v, w, x, y, z\}$, $\{r, s, t, u, v, w, z\}$, $\{r, s, t, u, v, w, x, y, z\}$;

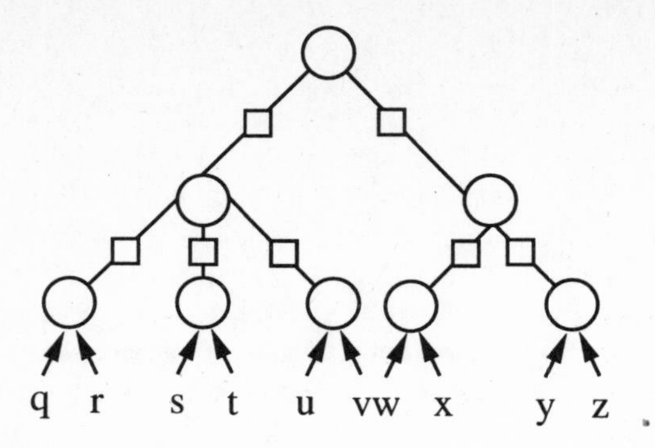

Figure 5.15 A worst case for searching a minimal sufficient subset. $\{q, t, v, x, z\}$ is minimal sufficient. Six pairs of propagations are required.

- second round: $\{s, t, w, z\}$, $\{s, t, w, x, y\}$, $\{t, w, x, y, z\}$, $\{s, w, x, y, z\}$, $\{r, t, u, v, w, x, y\}$, $\{r, t, u, v, w, z\}$, $\{r, s, t, u, v, w, x, y\}$, $\{r, s, u, v, w, z\}$;
- third round: $\{s, w, z\}$, $\{t, w, z\}$, $\{s, w, x, y\}$, $\{t, w, x, y\}$.

Altogether 18 pairs of propagation are required.

5.4.4 Redundancy and irrelevance

So far we have dealt with analysis of impact on a hypothesis, but *lack of impact* is also part of a sensitivity analysis.

Let h be a hypothesis, e the evidence, and let θ be a predefined threshold. A subset $e' \subseteq e$ is *redundant* if

$$\frac{P(h \mid e \setminus e')}{P(h \mid e)} > 1 - \theta.$$

If in the wet grass example we choose $\theta = 0.02$ then both e_W and e_J are redundant.

To investigate whether a finding x is redundant we need $P(h \mid e \setminus \{x\})$, and this is achieved through cautious propagation.

Although two findings s and t are redundant we cannot conclude that they can both be discarded (see, for example, the wet grass example). So, we must distinguish between redundance and irrelevance. A finding x is *irrelevant for* h if

$$\frac{P(h \mid e'' \setminus \{x\})}{P(h \mid e'')} > 1 - \theta \text{ for all } e'' \subseteq e,$$

where θ is a predefined threshold.

Note if x and y are irrelevant then

$$\frac{P(h \mid e'' \setminus \{x, y\})}{P(h \mid e'')} = \frac{P(h \mid e'' \setminus \{x, y\})}{P(h \mid e'' \setminus \{x\})} \frac{P(h \mid e'' \setminus \{x\})}{P(h \mid e'')} > (1 - \theta)^2$$

for all $e'' \subseteq e$.

An irrelevance analysis is much more involved than a redundance analysis, and we shall not go into it here.

5.5 Value of information

Whenever decisions under uncertainty are to be made, there is a quest for more information to reduce the uncertainty. However, information is seldom cost free, and there is, therefore, also a need for evaluating beforehand whether it is worthwhile consulting information sources. Furthermore, if several sources are available there is a need to come up with a strategy for a sequence of data requests.

5.5.1 Example

Consider the insemination example (Section 3.1.3 and Exercise 3.1). Before deciding on an action there is a possibility of acquiring information. A scan can be performed at a cost of 40 units, a blood test costs 10 units, a urine test costs 10 units, and a combined blood and urine test costs 15 units (see Fig. 5.16). Should any test be performed, and if so, in which order?

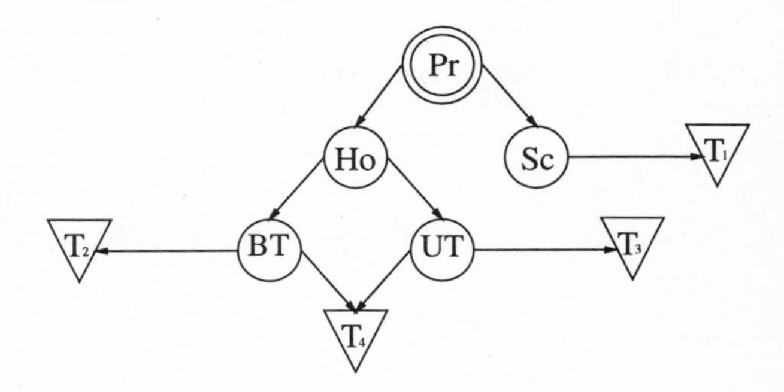

Figure 5.16 The data request situation in the example of Section 5.5.1. The double circled variable indicates that it is the driving hypothesis variable. A triangle indicates a test yielding the state of its parents.

First of all we shall attach a *value* to the various information scenarios. The driving force for evaluating an information scenario is the information on the hypothesis variable *Pr*. We therefore call this kind of data request situation *hypothesis driven*.

For our example, assume that six weeks after the insemination there are two possible actions, *na* (wait another six weeks) and *rp* (repeat the insemination). Assume also that Table 5.6 gives the outcome of each action given the state of *Pr*.

Such a table is called a utility table. Utilities need not be money, but they are numeric weights given to the various possible actions. We shall deal in more detail with utilities in Chapter 6.

Table 5.6 Utilities for the various combinations of actions and pregnancy situations: *na* means no action and *rp* means repeated insemination. The utility scale may be money, but it need not be.

	Actions	
Pr	*na*	*rp*
y	200	134
n	104	134

The value of any information scenario is a function V of the distribution of Pr. In our case, V is the *expected utility* of performing an optimal action.

$$V(P(Pr)) = \max_{a \in A} \sum_{h \in Pr} U(a,h)P(h).$$

A proper analysis of the data request situation should consist of an analysis of all possible sequences of tests (including the empty sequence). However, at this stage we shall limit ourselves to the *myopic* approach: if you are allowed to consult at most one information source, which option should be chosen?

The myopic approach is a proper limitation. Sometimes a single test does not yield anything by itself while its outcome may be crucial for selecting a second very informative test. We shall return to non-myopic data request in Section 5.5.3.

If test T with cost C_T yields the outcome t, then the value of the new information scenario is

$$V(P(Pr \mid t)) = \max_{a \in A} \sum_{h \in Pr} U(a,h)P(h \mid t).$$

Since the outcome of T is not known we can only calculate the *expected value*

$$EV(T) = \sum_{t \in T} V(P(Pr \mid t))P(t).$$

The *expected benefit* of performing test T is

$$EB(T) = EV(T) - V(P(Pr)).$$

The *expected profit* is

$$EP(T) = EB(T) - C_T.$$

The myopic data request task is to calculate the expected profit for the various tests and to choose the one with maximal expected profit (if positive).

To calculate the expected values in our example, one should determine $P(Pr)$, $P(UT)$, $P(BT)$, $P(Sc)$, $P(BT, UT)$ as well as $P(Pr \mid UT), \ldots, P(Pr \mid BT, UT)$. Now, instead of propagating all possible outcomes of the tests, Bayes' theorem

Table 5.7 $P(UT, BT \mid Pr = y)$.

	$UT = y$	$UT = n$
$BT = y$	0.51	0.13
$BT = n$	0.22	0.14

will yield the necessary probabilities through one propagation only ("y" means "$Pr = y$"):

$$P(y \mid UT) = \frac{P(UT \mid y)P(y)}{P(UT)}$$
$$P(n \mid UT) = 1 - P(y \mid UT),$$

where $P(UT \mid y), \ldots, P(Sc \mid y)$ are obtained by entering $Pr = y$ as a finding to the network.

There is a problem remaining for the combined test $T_4 = BT\&UT$. To calculate $EB(BT\&UT)$ we need both $P(BT, UT)$ and $P(BT, UT \mid Pr = y)$, which cannot be achieved directly from a propagation in the network. Propagation of variables can be used here.

In Table 5.7 the probabilities for $P(UT, BT \mid Pr = y)$ in the insemination example is given. The expected benefit from a combined urine and blood test is calculated to be zero, while scanning has a positive expected benefit (see Exercise 5.13).

5.5.2 Hypothesis driven data request

Hypothesis driven data request is characterized by the evaluation of the state of the network which is based only on the probability distribution of a hypothesis. To be precise: H is a *hypothesis* with a set of mutually exclusive states $h_1, \ldots, h_n$ and V is a *value function* $V(P(H)) : [0; 1]^n \rightarrow R$.

The *expected value* of performing a test T is

$$EV(T) = \sum_{t \in T} V(P(H \mid t))P(t).$$

The *expected benefit* is $EV(T) - V(P(H))$.

In principle, any value function may be used. However, a particular class of functions, namely *convex functions*, are best suited.

Definition. A function $f : R^n \rightarrow R$ is convex if for any two points P_1, P_2 on the graph of f, the line segment P_1P_2 lies above the graph (see Fig. 5.17). Mathematically the property is expressed as follows:

$$\forall t \in [0, 1], \forall \underline{x}, \underline{y} \in R^n : tf(\underline{x}) + (1-t)f(\underline{y}) \geq f(t\underline{x} + (1-t)\underline{y}).$$

The reason why a convex function is well suited is due to the following.

Theorem 5.4 *If the value function is a convex function then the expected benefit of performing a test is never negative.*

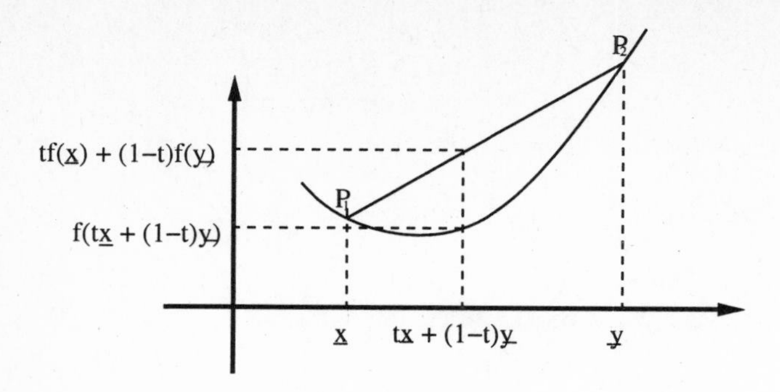

Figure 5.17 A convex function. The line segment between two points of the graph lies above the graph.

(Proof in Appendix B.)

The following value functions are all convex (proofs in Appendix B), and therefore they have the property of Theorem 5.4.

Utilities

In the example in Section 5.5.1, a utility based value function was used:

$$V(P(H)) = \max_{a \in A} \sum_{h \in H} U(a, h)P(h),$$

where A is a set of mutually exclusive non-intervening actions, and U is a utility table for the actions given the hypothesis. An action is non-intervening if it does not change the state of a variable in the network (see Ch. 6).

Behind the use of utility based value functions lies the assumption that an action with highest expected utility is taken. This means that in any information scenario you have an opinion on which action to take. Therefore you would only take the bother to acquire new information if this may change your mind. Actually it is so that:

Theorem 5.5 *The expected benefit of performing test T is zero if and only if the optimal action is unchanged no matter the outcome of T.*

The proof is given in Appendix B.

Notice (see Exercise 5.14) that the action of selecting the state of H with highest probability can be modelled by utilities yielding a value function equivalent to

$$V(P(H)) = \max_{h \in H} P(h)$$

and $EV(T)$ is then positive if and only if some outcome of T causes a new state to have highest probabilities.

Entropy

If there is no proper model for actions and utilities, the reason for acquiring more information is to decrease the uncertainty of the hypothesis. This means that you will give high values to probabilities close to zero and one, while probabilities in the middle area should have low values. A classical function with this property is *entropy*.

The formula for entropy of a distribution over H is

$$ENT(P(H)) = -\sum_{h \in H} P(h)\log_2(P(h)),$$

where $p\log_2 p = 0$ if p vanishes.

Entropy is a measure of how much the probability mass is scattered around on the states. If the distribution is even, and the number of states in H is n, then the entropy is $\log_2 n$. In the limit, if the probability mass is concentrated in one state, the entropy is zero. As we want the value function to increase with preference, we let an entropy based value function be

$$\begin{aligned} V(P(H)) &= -ENT(P(H)) \\ &= \sum_{h \in H} P(h)\log_2(P(h)). \end{aligned}$$

Weight of evidence

If the hypothesis has only two states h and $\overline{h}$, we can use the ratio between $P(h)$ and $P(\overline{h})$ as a value function. For convenience we take the log:

$$WOE(P(H)) = \log P(h) - \log P(\overline{h}) = \log P(h) - \log(1 - P(h)).$$

If evidence e is acquired, the difference in WOE is

$$WOE(P(H \mid e)) - WOE(P(H)) = \log\frac{P(h \mid e)P(\overline{h})}{P(\overline{h} \mid e)P(h)}.$$

By Bayes' rule we get

$$WOE(P(H \mid e)) - WOE(P(H)) = \log\frac{P(e \mid h)}{P(e \mid \overline{h})}.$$

This is the log of the ratio of the likelihoods for h and $\overline{h}$. In other words, the difference in WOE is a measure of the weight on h from e.

Warning. WOE is convex only for $P(h) > \frac{1}{2}$(see Appendix B and Exercise 5.12). If you take the absolute value of WOE you get a convex value function, but then it does not really measure weight of evidence if the evidence changes h from being most probable to least probable.

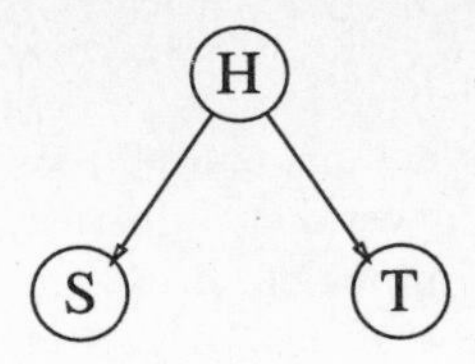

Figure 5.18 The causal relations between disorder H and tests S and T in the non-myopic example.

Variance

If the states of H are numeric, another classical measure can be used, namely the *variance*. Again, since small variances are preferred, the value function becomes

$$V(P(H)) = -\sum_{h \in H} (h - \mu)^2 P(h),$$

where $\mu = \sum_{h \in H} h P(h)$.

It is up to the modeller to specify the value function. If decisions with known utilities are attached to the hypothesis variable, then the utility value function should be preferred. If this is not the case, the user will mainly be interested in the precision of a diagnosis.

In, for example, the case of a Boolean hypothesis with states 0 and 1, the entropy function is $\log p^p(1-p)^{1-p}$ and the variance function is $-p(1-p)$. These two functions reflect that the value of p increases as it approaches its bounds 0 and 1. The entropy function is rather drastic in the way that the slope is infinite for 0 and 1. Therefore, small changes of p close to 0 and 1 will be highly valued. On the other hand, the variance is of polynomial degree 2, and the slope close to the bounds is 1 and -1, giving changes almost even value no matter how close to the bound.

Finally, note that linear value functions are useless. (Proof in Appendix B.)

5.5.3 Non-myopic data request

It may happen that no test has a positive expected profit, while the expected benefit for a pair of tests is greater than the costs of them. In that case the myopic approach is misleading. Consider the following simple example.

A disorder H with prior probability 0.02 has two tests T and S each with a 10% risk of giving false positive as well as false negatives. The value function is $\max_{h \in H} P(H)$ (the one originating from selecting the state with highest probability as the diagnosis). The causal relations are given in Figure 5.18.

Initially, $V(P(H)) = 0.98$ with *no* as the chosen diagnosis. According to Theorem 5.5 a test will have a positive expected benefit only if some test result gives *yes* highest probability. This may only happen for test result *yes*.

$$P(H = yes \mid T = yes) = 0.16,$$

however. So no single test will have a positive expected benefit.

On the other hand, $P(H = yes \mid T = yes, S = yes) = 0.62$ which means that the expected benefit from performing both tests is positive. This highlights that the value of performing a test is not only the change of value; performing a test has the side effect that it changes the expected benefit from performing other tests. Therefore, the options in this example are:
T and ?S: observe T and decide afterwards whether to observe S;
S and ?T: observe S and decide afterwards whether to observe T;
T and S: observe T and S.

Unless costs may be saved by performing both tests simultaneously (laboratory discount), **T and S** can never be better than the first two options. We analyze **T and ?S**.

The expected profit from observing T is

$$EP(T) = \sum_{t \in T} V(P(H \mid t))P(t) - Cost(T) - V(P(H)).$$

If the test result is t then the expected profit of observing S is

$$EP(S \mid t) = \sum_{s \in S} V(P(H \mid s, t))P(s \mid t) - Cost(S) - V(P(H \mid t)).$$

If $EP(S \mid t) \leq 0$ then S is not performed. Hence

$$EP(T \ and \ ?S) = EP(T) + \sum_{t \in T} P(t) \max(0, EP(S \mid t)).$$

To be able to calculate $EP(T \ and \ ?S)$ we (in addition to the probabilities from myopic data request) also need $P(H \mid S, T)$ and $P(S \mid T)$, and in an analysis of **S and ?T**, we would also need $P(T \mid S)$.

Since

$$P(S, T) = \sum_{h \in H} P(S, T \mid H)P(H),$$

we can by Bayes' formula conclude that if $P(S, T \mid H)$ is available, all sufficient information will be present. Propagation of variables can give us $P(S, T \mid H)$.

In our example it is even easier: S and T are independent given H. Then $P(T, S \mid H) = P(T \mid H)P(S \mid H)$ and the analysis can be performed through propagation of the states of H.

By setting costs for each test to 0.005 we get $EP(T \ and \ ?S) = 0.0014$. To compare, we have $EP(T \ and \ S) = -0.0037$.

5.6 Pathfinder

Pathfinder is a normative expert system that assists community pathologists with the diagnosis of lymph-node pathology. It is integrated in the commercial system *Intellipath* which also includes video discs. Intellipath is used by practicing pathologists and by pathologists in training as a teaching and learning tool.

Pathfinder consists of a Bayesian network with one hypothesis variable, *Disease*, which has more than 60 states covering more than 60 mutually exclusive diseases.

This means that the system does not diagnose multiple diseases. There are 130 information variables and no mediating variables. The network is structured so that each information variable is a child of *Disease*. In addition, there are relations between the information variables (see Figure 5.19). The maximal number of parents for a variable is six.

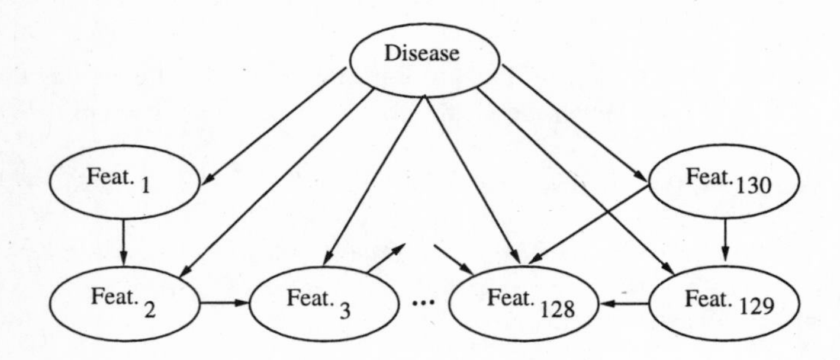

Figure 5.19 The structure of Pathfinder. The *Disease* variable is a parent of all feature variables.

Pathfinder has a facility which recommends to the user which test to perform next. It is a myopic value of information facility with a utility table for the value function.

A utility table $U(Disease, Disease)$ has been established. For each pair of diseases (d_i, d_j) a utility value is given to the situation where the patient suffers from d_i, but she is diagnosed as having d_j and treated accordingly. The utilities used are measured in the unit *micromort*, which is the probability 10^{-6} of sudden painless death. The costs of the tests – including money as well as discomfort for the patient and risks of side effets – are also measured in micromort. The principle behind the transformation of money to the micromort scale is to ask the patient how much she is willing to pay to decrease a risk of sudden painless death with 10^{-6} (provided the risk is small already).

Naturally, the acquired numbers $U(d_i, d_j)$ are very uncertain, and therefore Pathfinder does not use the utility table to recommend the disease with maximal expected utility as a diagnosis decision (see Ch. 6). The uncertainty on the utilities is not as serious when they are used for guiding information collection. To investigate how sensitive Pathfinder's recommendations are to the utility table, an alternative table has been tried out. It has $U(d_i, d_j) = 1$ when both diseases are benign or both are malignant, and $U(d_i, d_j) = 0$ otherwise. The recommendations often turned out to be similar.

A study has been performed where Pathfinder was compared to a simple Bayes version and to an expert's conclusions. The results show that the diagnostic accuracy of Pathfinder is greater than that of the simple Bayes version, and it is just as good as the experts.

5.7 Bibliographical notes

Variable propagation is an idea of Xu (1994). Max-propagation and insertion for fast retraction are proposed by Dawid (1992) and Cowell & Dawid (1992). Cautious propagation is described in Jensen (1995b); it is a modification of the propagation method suggested by Shafer & Shenoy (1990). A measure for calculating data conflict was first proposed by Habbema (1976). The measure presented here is due to Jensen et al. (1991). The approach has been extended by Kim & Valtorta (1995). Sensitivity analysis is part of *explanation*. It was systematically studied by Suermondt (1992), and in Madigan & Mosurski (1993) an implementation of various facilities is reported. Value of information is formally treated in Howard (1966) and Lindley (1971), where utilities are guiding the test selection. The myopic approximation was introduced by Gorry & Barnett (1968). In Ben-Bassat (1978) entropy and variance is used. Glasziou & Hilden (1989) discuss quasi-utilities like entropy, variance and weight of evidence. Pathfinder is described in Heckerman et al. (1992) and Heckerman & Nathwani (1992a,b).

Exercises

Exercise 5.1 ^{H} Answer the questions in Section 5.1 on *transmission of symbol strings.*

Exercise 5.2 ^{H} Answer the questions in Section 5.1 on *stud farm.*

Exercise 5.3 In the junction tree below, $P(A, C, G, I)$ shall be calculated.

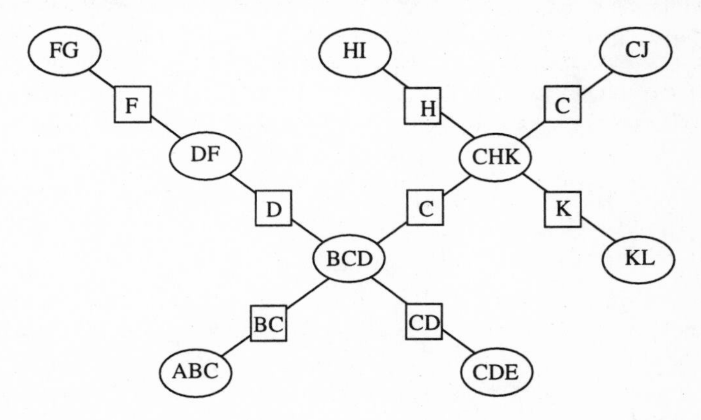

Figure for Exercise 5.3.

(i) Determine the control structure for firing of variables as well as for propagation of variables. Compare the complexity of the two methods.

(ii) Assume that all variables are binary. Determine a root V for

$$CollectEvidence(V, (A, C, G, I))$$

such that the resulting tables are of minimal size.

Exercise 5.4 In *transmission of symbol strings* (Section 3.2.3) the sequence **baaca** is received. Determine the most probable transmitted word.

Exercise 5.5 ^{H} Let John be sick in the *stud farm* example from Section 3.2.1 (Fig. 3.9.)

(i) Determine the most probable configuration of *aA* and *AA* for the remaining horses.

(ii) Do the same for the model in Exercise 3.3(i).

Exercise 5.6 Consider the junction tree from Exercise 5.3, and assume that evidence has been entered on the variables A, G, I, J and K.

(i) The clique CDE is used as the root for a HUGIN propagation. Determine the sets of evidence for which the probabilities can be accessed directly.

(ii) Which sets of evidence can further be accessed through a cautious propagation?

(iii) Compare the number of table multiplications for the two propagations.

Exercise 5.7 ^{H} Consider the case in Exercise 3.6(iii).

(i) What is the global conflict?

(ii) Does the change of probabilities of *parental error* explain the conflict?

(iii) Is there a red herring among the findings?

(iv) The task of the BOBLO network is to analyze coherence between the stated parents and the reported factors. Discuss whether conflict analysis could do the work (possibly on another network).

Exercise 5.8 Prove Proposition 5.2.

Exercise 5.9 ^{H} To the network from Exercise 3.15 for scene analysis, the image processor reports to have identified a big plate, a coffee cup, a coffee pot, a jar with orange contents, a knife and a fork. Perform a conflict analysis.

Exercise 5.10 ^{H} Perform a sensitivity analysis of the case in Exercise 3.6(iii).

Exercise 5.11 ^{H} Perform a sensitivity analysis of the case in Exercise 5.9.

Exercise 5.12 The hypothesis H has two states h and $\overline{h}$, and $P(H) = (0.493, 0.507)$. The binary variable B can be observed, and we have $P(B) = (0.7, 0.3)$, $P(h \mid b_1) = 0.7$, and $P(h \mid b_2) = 0.01$.

Use the value function *WOE*, and show that the expected benefit of observing B is negative.

Exercise 5.13 ^{H} In the Insemination example (Section 5.5.1), calculate the expected benefit of *BT* & *UT* and *Sc*.

Exercise 5.14 Let the hypothesis variable H have n states. Introduce an action variable A with the same states as H; let the utility table be as follows:

$$U(h, a) = \begin{cases} 1 & \text{if } h \text{ and } a \text{ are the same} \\ 0 & \text{otherwise.} \end{cases}$$

Show that a value function based on U corresponds to selecting a hypothesis state of highest probability.

Chapter 6

Actions

A Bayesian network serves as a model for a part of the world, and the relations in the model reflect causal impact between events. The reason for building these computer models is to use them when taking decisions. That is, the probabilities provided by the network are used to support some kind of decision making. In principle there are two kinds of decisions, namely *test-decisions* and *action-decisions*.

A test-decision is a decision to look for more evidence to be entered into the model, and an action-decision is a decision to change the state of the world. In real life this distinction is not very sharp; tests may have side effects, and by performing a treatment against a disease, evidence on the diagnosis may be acquired. In order to be precise we should say that decisions have two *aspects*, namely a test aspect and an action aspect. The two aspects are handled differently in connection with Bayesian networks, and accordingly we treat them separately. In Section 5.5 we presented methods for test decisions and in this chapter we deal with actions.

Actions should also be divided into two types, namely *intervening actions* which force a change of state for some variables in the model, and *non-intervening actions* of which the impact is not a part of the model.

Although both observations and intervening actions change the probability distributions in the model, they are fundamentally different. To highlight this, consider the example in Figure 6.1.

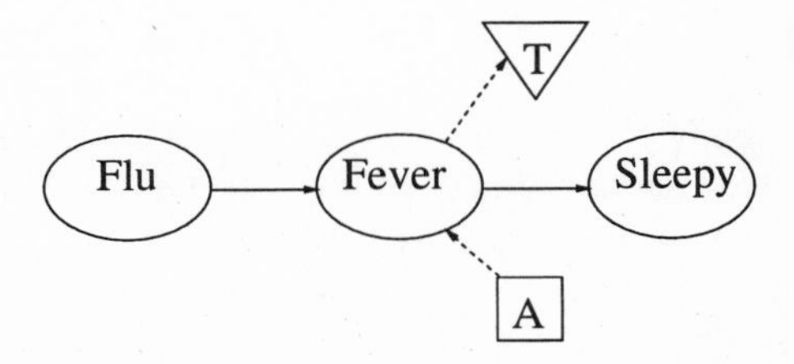

Figure 6.1 A simple flu decision model with an action (aspirin) and a test (temperature) attached. The action has no impact on $P(Flu)$.

If T yields a temperature of 37°C (99°F), both the probability for *Flu* and for *Sleepy* are decreased. On the other hand, if the fever is set to 37°C by taking aspirin, then it does not cure the flu (though some people seem to believe so), but still the probability for *Sleepy* is reduced due to the reduced fever. That is, *impact of intervening actions can only follow the direction of the causal links.*

The example stresses an important point concerning the use of Bayesian networks. Using Bayes' theorem it is easy to establish the model in Figure 6.2 which reflects diagnostic reasoning from symptoms to disease.

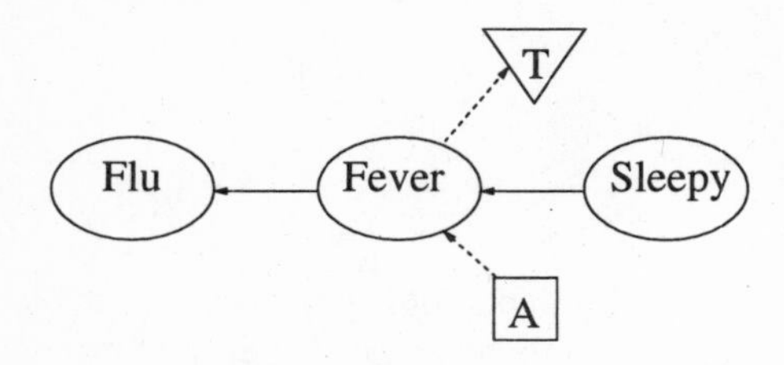

Figure 6.2 A decision model with a Bayesian network equivalent to the one in Figure 6.1; however, the action will, in the model, influence $P(Flu)$. To be correct an arrow from A to *Flu* must be added.

From the point of view of entering evidence and propagating probabilities the two Bayesian networks in Figure 6.1 and Figure 6.2 are equivalent. So, why bother with emphasizing that the links in the network should be causal links? The difference becomes apparent when taking an aspirin: in Figure 6.2, taking an aspirin will cure the flu but will have no effect on sleepiness.

In Section 6.1 we discuss the concept of utilities through two examples. Section 6.2 shows how to use Bayesian networks for decisions on non-intervening actions, and Section 6.3 treats intervening actions. The methods in both Section 6.2 and Section 6.3 deal with situations where you have to decide on one out of a set of possible actions. In Section 6.4 we consider decision problems for sequences of possible actions.

6.1 Utilities

We shall treat decision problems in a certain framework, namely *utility theory*. Decisions are taken because they may be of use in some way. Therefore, the various decisions should be evaluated on the basis of the usefulness of their consequences. We shall assume that "usefulness" is measured on a numerical scale called a *utility scale*, and if several kinds of utilities are involved in the same decision problem, then the scales have a common unit. This assumption may seem dubious; it is treated in the extensive literature on utility theory, e.g. Lindley (1971) and Winterfeldt & Edwards (1986). We shall touch on one of the general methods in Section 6.1.1.

The utility of an action may depend on the state of some variables called *determining variables*. For example, the utility of a treatment with penicillin is

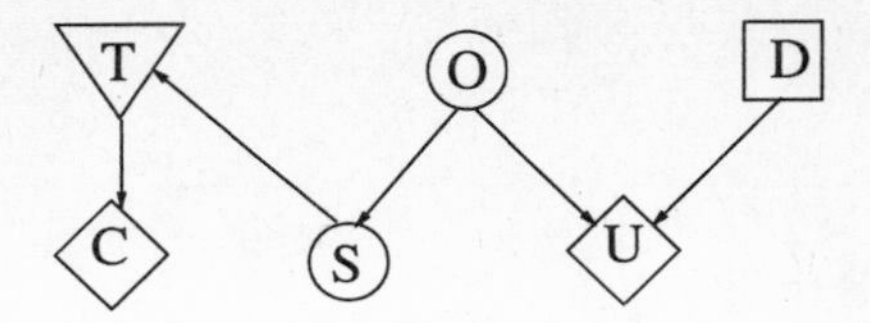

Figure 6.3 A graphical representation of the oil wildcatter's problem. U and C are utility nodes representing the outcome of drilling and the costs of testing. The arrow from S to T indicates that the test results are findings on S.

dependent on the type of infection and whether the patient is allergic to penicillin. Let $A = (a_1, \ldots, a_n)$ be a set of mutually exclusive actions, and let H and G be determining variables. What is required in order to specify the problem of deciding between the actions in A is a *utility table* $U(A, H, G)$ yielding the utility for each configuration of action and determining variables.

To give some insight into the considerations and calculations with utilities, we shall go through two examples. In the sections to come, most issues are dealt with in a more systematic manner.

6.1.1 The oil wildcatter

> An oil wildcatter must decide either to drill or not to drill. He is uncertain whether the hole is dry, wet or soaking. The wildcatter could take seismic soundings that will help determine the geological structure of the site. The soundings will give a closed reflection pattern (indication of much oil), an open pattern (indication of some oil) or a diffuse pattern (almost no hope of oil).

The wildcatter has two decisions to make, namely whether to test with seismic soundings (T, costing K$10) or whether to drill (D, costing K$70). D is determined by the state of the hole (O) which can be either *dr, wt* or *sk*. If we measure utility in K$ then a table giving the outcomes of drilling given the state of O must be provided. The test provides the state of the variable S with states *cl, op* and *di*. The relation between O and S is that O has a causal effect on S. In Figure 6.3 a graphical representation of the decision problem is given, and the associated numerical specification is listed in Table 6.1.

Since the wildcatter does not know the state of O, he has to use the probability distribution $P(O)$. His expected utility from not drilling is zero no matter the state of O. That is, $EU(\textit{not drill}) = 0$. If he drills there is 50% risk of loosing K$70; there is a 30% chance of winning K$50, and a 20% chance of winning K$200. His *expected utility* is the weighted average of these outcomes, i.e.

$$EU(drill) = \sum_{O} U(drill, O)P(O) = 20.$$

Table 6.1 Probability and utility tables for the oil wildcatter's problem.

	$O = dr$	$O = wt$	$O = sk$
$S = cl$	0.1	0.3	0.5
$S = op$	0.3	0.4	0.4
$S = di$	0.6	0.3	0.1

$P(S \mid O)$

dr	0.5
wt	0.3
sk	0.2

$P(O)$

	$O = dr$	$O = wt$	$O = sk$
$D = drill$	−70	50	200
$D = not\ drill$	0	0	0

$U(D, O)$

We expect the wildcatter to act in order to maximize the expected utility. In this situation he will drill, and the maximum expected utility is

$$MEU(D \mid not\ test) = \max(EU(drill), EU(not\ drill)) = 20.$$

A short digression should be made here. In the above specification of the problem there is a hidden assumption about the wildcatter, namely that he is willing to risk a loss of K\$70 for a chance of winning K\$200, as long as the expected outcome is positive. Such a behaviour is called *risk-neutral*. Some people will not take that risk unless the expected outcome is well above zero. They are called *risk-averse*. Most people are risk-averse when the maximal loss is above a certain level. This is why insurance companies make money. However, we often act *risk-prone* when the loss may be small and the maximal gain is high. This is why lotteries produce a profit. If our wildcatter had not been risk-neutral the money scale could not be used as a utility scale. We would then have had to make a non-linear transformation of the money scale to his private utility scale. We shall not go deeper into this, but assume that our decision makers are risk-neutral with respect to the utility scales used.

Back to the wildcatter's problem. Should he perform a test or should he go ahead drilling? Without performing any sophisticated calculations we can say that *the extra cost of testing can only be justified if some outcome of the test makes him change his mind.*

So, may an outcome of the test prevent him from drilling? The worst test result for the wildcatter would be that S is in the state di. Now $P(O \mid S = di)$ can be calculated in the usual way by using Bayes' rule. The result is

$$P(O \mid di) = (0.732, 0.22, 0.049).$$

This gives

Table 6.2 Expected utilities of actions given test outcomes. The numbers in boldface are the expected utilities achieved by taking an optimal action.

	$S = cl$	$S = op$	$S = di$
$D = drill$	**87.5**	**32.9**	−30.5
$D = not\ drill$	0	0	**0**

$$EU(drill \mid di) = \sum_{O} U(drill \mid O)P(O \mid di) = -30.5.$$

This result says that it may be worth while performing the test since – in the case of the test result *di* – the initial decision to drill will be altered. However, it does not tell us whether the test is altogether worth its cost. In Table 6.2 we give an overview of action and expected utility given the various test results. If we assume that the wildcatter will choose an action with highest expected utility, we can get a direct relation between test result and expected utility by taking the maximum of each column in Table 6.2:

$$MEU(D \mid S, test) = (87.5, 32.9, 0).$$

We can now calculate the maximal expected utility from performing the test by averaging $MEU(D \mid S, test)$ over the probabilities of the test results ($P(S)$) and subtract the cost of T:

$$\begin{aligned} MEU(D \mid test) &= \sum_{S} MEU(D \mid S, test)P(S) - C \\ &= (87.5 \cdot 0.24 + 32.9 \cdot 0.35 + 0) - 10 = 22.2. \end{aligned}$$

Since $MEU(D \mid test) > MEU(D \mid not\ test)$ it pays to perform the test. Test decisions are also treated in more depth in Section 5.5.

6.1.2 Management of effort

In your computer science studies you attend two courses, they are *Graph Algorithms* and *Decision Support Systems*. In the middle of the term you realize that you can't keep pace. You can either reduce your effort in both courses slightly or you can decide to attend one of the courses superficially only. What is the best decision?

You have three possible actions:

Gd keep pace in Graph Algorithms and follow Decision Support Systems superficially;

SB slow down in both courses;

Table 6.3 The conditional probabilities of the final marks in Graph Algorithms *(GA)* and Decision Support Systems *(DSS)* given the efforts *keep pace (kp)*, *slow down (sd)*, and *follow superficially (fs)*.

	kp	*sd*	*fs*
0	0	0	0.1
1	0.1	0.2	0.1
2	0.1	0.1	0.4
3	0.2	0.4	0.2
4	0.4	0.2	0.2
5	0.2	0.1	0

$P(GA \mid effort)$

	kp	*sd*	*fs*
0	0	0	0.1
1	0	0.1	0.2
2	0.1	0.2	0.2
3	0.2	0.2	0.3
4	0.4	0.4	0.2
5	0.3	0.1	0

$P(DSS \mid effort)$

Dg keep pace in Decision Support Systems and follow Graph Algorithms superficially.

The results of the actions are your final marks for the courses (many students look at it like that). The marks are integers between 0 and 5, where 0 and 1 are failing marks. You have certain expectations for the marks given your effort in the rest of the term. They are shown in Table 6.3.

A way of solving your decision problem would be to say that the numeric value of the mark is a utility, and you want to maximize the sum of the expected marks. The calculations would then be

$$EU(Gd) = \sum_{m \in GA} P(m \mid kp)m + \sum_{m \in DSS} P(m \mid fs)m = 3.5 + 2.3 = 5.8$$

$$EU(SB) = \sum_{m \in GA} P(m \mid sd)m + \sum_{m \in DSS} P(m \mid sd)m = 2.9 + 3.2 = 6.1$$

$$EU(Dg) = \sum_{m \in GA} P(m \mid fs)m + \sum_{m \in DSS} P(m \mid kp)m = 2.3 + 3.9 = 6.2.$$

From this you would conclude that you should follow Graph Algorithms superficially but keep pace in Decision Support Systems.

However, do the marks really reflect your utilities? If, for example, you had the same number of marks, but the numeric values were 0, 5, 6, 8, 9, 10 you would have come to another conclusion. The problem is that you cannot expect that a difference of 1 in mark number always represents the same difference in utility. Actually, in this case your subjective utility is not increasing in the numeric value of the mark: at your university it is so that if you fail you are given another chance, but if you pass you are not allowed to try again to get a better mark. Therefore you find that the worst mark to get is a 2 rather than a 0!

To overcome this problem, the mark scale is mapped into a utility scale ranging from 0 to 1. The best possible mark (5) is given the utility 1, and the worst possible mark (2) gets the utility 0.

Table 6.4 Utilities for the various marks (the same for both courses).

Mark	0	1	2	3	4	5
Utility	0.05	0.1	0	0.6	0.8	1

The intermediate marks are given utilities by imagining that you have a choice between two games.

Game 1: you get for certain the mark x.

Game 2: you get mark 5 with probability p, and you get 2 with probability $1 - p$.

Which game would you prefer?

If $p = 0$ you would prefer Game 1, and for $p = 1$ Game 2 would be best. For some p between 0 and 1 you would be indifferent, and *this p is the utility for the mark x.*

In Table 6.4 we have performed the utility assessment for you. The utilities assessed are only for one course. We will now assume that the utility of marks for several courses is the sum of the individual utilities. Note that this is not evident (it might, for example, be that you prefer two 2s to failing both courses, which would delay your studies considerably).

In Figure 6.4 the decision model is illustrated.

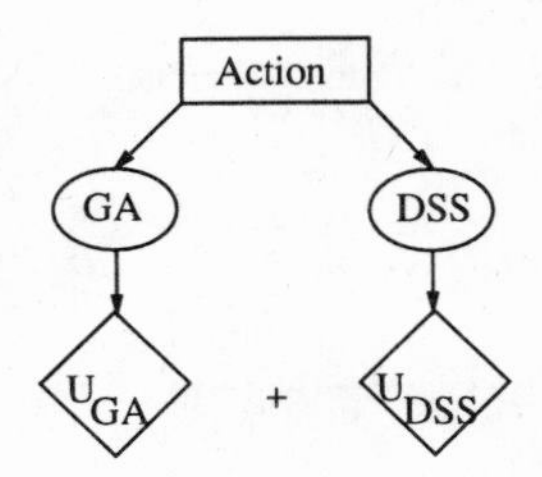

Figure 6.4 A decision model for effort.

To find the optimal decision the calculations are

$$EU(action) = \sum_{m \in GA} P(m \mid action) U_{GA}(m) + \sum_{m \in DSS} P(m \mid action) U_{DSS}(m).$$

We get $EU(Gd) = 1.015$, $EU(SB) = 1.07$, $EU(Dg) = 1.035$ and the optimal action is SB.

6.2 Non-intervening actions

To make things easy we start with the following situation.

- We have a set $a_1, \ldots, a_n$ of mutually exclusive actions and one determining variable H in a Bayesian network.
- The actions have no impact on $P(H)$ (see Fig. 6.5).
- For each state h of H and each action a we have a number $U(a, h)$ describing the utility of action a if H is in state h.

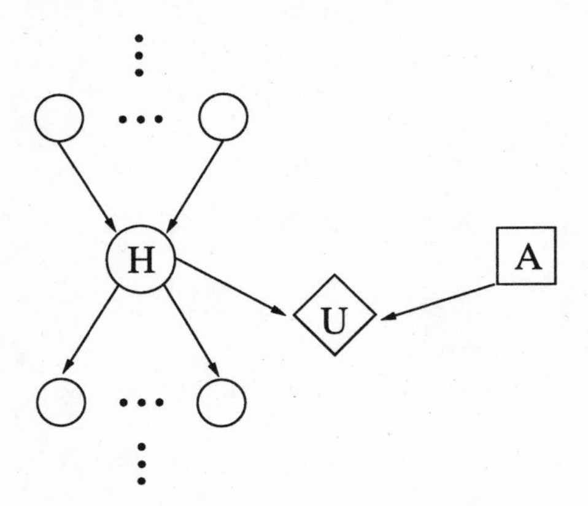

Figure 6.5 Action and utilities based on one determining variable in a Bayesian network.

The way to find out which action to take is to use the utility table $U(a, h)$ and choose the one with highest utility.

However, usually we only have a probability distribution $P(H)$. In the case of non-intervening actions where $P(h \mid a) = P(h)$, the *expected utility* from taking action a is

$$EU(a) = \sum_{h \in H} U(a, h)P(h).$$

The *optimal action* is $\text{Opt}(A) = \arg\max EU(a)$, where "arg max" means the "argument which maximizes".

The *maximal expected utility* is $MEU(A) = \max EU(a)$.

Notation. We shall sometimes use the notation $EU(a \mid e)$, $\text{Opt}(A \mid e)$, etc., where e is some evidence. This means the expected utility given e, etc. We shall also use $EU(A \mid B)$, where B is a variable. This is a notation for a table of $EU(A \mid b)$s.

Insemination. As an example, consider the insemination example, Section 3.1.3 (the conditional probabilities are listed in Exercise 3.1). Suppose that six weeks after the insemination there are the two actions *na* (wait another six weeks) and *rp* (repeat the insemination).

Let the profit of a successful pregnancy be 200 units, the price of keeping the cow non-pregnant for an extra six weeks is 30 units and the price of *rp* is 10 units.

So, the utility of *na* given $Pr = y$ is 200 and if $Pr = n$ we cannot hope for more than $200p - 2 \cdot 30 - 10$, where $p = 0.87$ is the success rate of inseminations

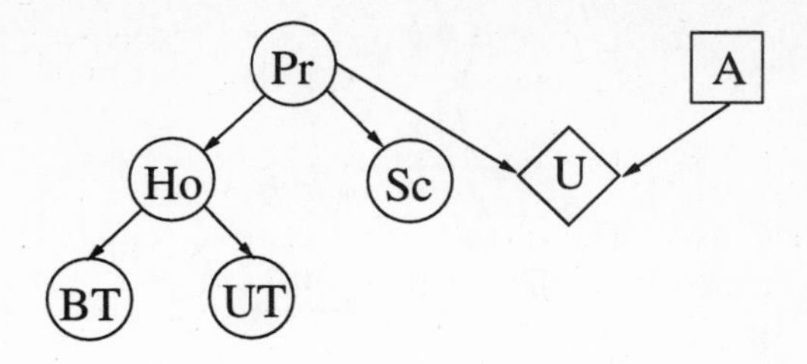

Figure 6.6 Insemination with an action node and a utility node.

Table 6.5 Utility table $U(A, Pr)$ for insemination. We assume that the action rp involves an interruption of a possible pregnancy.

	na	rp
y	200	134
n	104	134

(this is under the assumption that the cow is taken out of production if the second insemination also fails). The utilities for rp is $200p - 30 - 10$ regardless of the state of Pr (see Table 6.5).

Now, six weeks have gone after the insemination. If we have no further information, then $P(Pr) = (0.87, 0.13)$, and

$$EU(na) = 200 \cdot 0.87 + 104 \cdot 0.13 = 188$$

$$EU(rp) = 134.$$

So, $\mathrm{Opt}(A) = na$ (no surprise).

Next we get a negative blood and urine test. This reduces the probability of pregnancy to 0.53 and now we have

$$EU(na \mid BT = n, UT = n) = 200 \cdot 0.53 + 104 \cdot 0.47 = 155$$

$$EU(rp \mid BT = n, UT = n) = 134$$

and still na should be preferred.

A simple analysis of this example shows that if $P(Pr = y) < 0.31$ then the insemination should be repeated. However, if the determining variable has more than two states the situation is not as easy.

A calculation trick. Through a couple of tricks it is possible to have the above calculations performed in a Bayesian network.

Table 6.6
Normalized utilities for pregnancy.

	na	*rp*
y	1	0.31
n	0	0.31

- Perform a linear transformation of the utility table such that the minimal utility is zero and the maximal utility is one. Call this normalized table $NU(A, Pr)$ (see Table 6.6).
- Add two variables A (with states *na* and *rp*) and U (with states y and n).
- Let U be a child of A and Pr (see Fig. 6.6).
- Give $P(A)$ the even distribution (0.5, 0.5).
- Give $P(U = y \mid A, Pr)$ the table $NU(A, Pr)$.

Now, after evidence has been entered into the network, the expected normalized utility of performing action a can be read from the network by entering a as a finding. The resulting probability of $U = y$ is then the expected normalized utility, $ENU(a \mid e)$.

Proof. (We skip the conditioning on the evidence, e.)

$$\begin{aligned} ENU(a) &= \sum_{h \in Pr} NU(a, h) P(h \mid a) \\ &= \sum_{h \in Pr} P(U = y \mid a, h) P(h \mid a). \\ &= \sum_{h \in Pr} P(U = y, h \mid a) \\ &= P(U = y \mid a). \end{aligned}$$

The linear transformation from $U(A, Pr)$ to $NU(A, Pr)$ is of the form

$$NU(A, Pr) = \alpha U(A, Pr) + \beta. \tag{6.1}$$

Therefore, the optimal action is also the one which maximizes the expected normalized utility.

So

$$\text{Opt}(A) = \arg\max ENU(a).$$

This is not all: instead of entering all possible actions it is enough to enter $U = y$ as evidence. Then A is given a probability distribution, and *the state with highest probability is the optimal action*, and $\text{Opt}(A)$ is obtained through one propagation only.

Proof.

$$\begin{aligned}
\text{Opt}(A) &= \arg\max ENU(a) \\
&= \arg\max P(U = y \mid a) \\
&= \arg\max \left(\frac{P(a \mid U = y)P(U = y)}{P(a)} \right) \\
&= \arg\max \left(\frac{P(a \mid U = y)}{0.5} \right) \\
&= \arg\max P(a \mid U = y).
\end{aligned}$$

Note. $P(a)$ above is in fact $P(a \mid e)$, but since A is independent of H given any evidence then $P(a \mid e) = P(a)$.

For later use, let us see how it may be possible to calculate the actual value of the expected utilities from the probabilities in the network.

From (6.1) we get

$$EU(A \mid e) = \frac{1}{\alpha}(ENU(A \mid e) - \beta),$$

where

$$ENU(A \mid e) = P(U = y \mid A, e) = \frac{P(A \mid U = y, e)P(U = y \mid e)}{P(A)}.$$

What remains is therefore $P(U = y \mid e)$, but this can be read from the network before $U = y$ is entered.

6.2.1 Several sets of non-intervening actions

Suppose we have two action variables A_1 and A_2 with determining variables H_1 and H_2. The utility function U is then a function of A_1, A_2, H_1 and H_2, and the decision problem can be solved by treating pairs $(a_1, a_2) \in A_1 \times A_2$ as one action.

This approach has a computational complexity which grows exponentially with the number of action variables. Sometimes the utility function is decomposed as for example the sum of two local utility functions U_1 and U_2 (see Fig. 6.7).

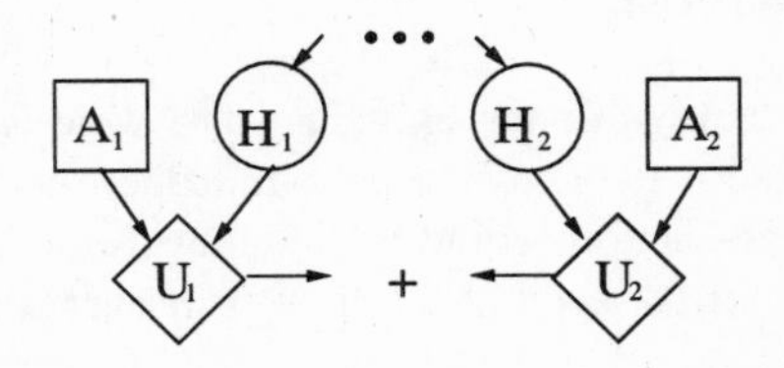

Figure 6.7 A decision problem with the utility function being a sum of two local utilities.

Then the decision problem can be solved by solving the two local decision problems separately.

Theorem 6.1 *Let A_1 and A_2 be nonintervening action variables with hypothesis variables H_1 and H_2, and let $U(A_1, A_2, H_1, H_2)$ be the utility function.*
If $U(A_1, A_2, H_1, H_2) = U_1(A_1, H_1) + U_2(A_2, H_2)$ then
(i) $MEU(A_1, A_2) = MEU(A_1) + MEU(A_2)$
(ii) $\text{Opt}(A_1, A_2) = (\text{Opt}(A_1), \text{Opt}(A_2))$.

Proof.

$$
\begin{aligned}
MEU(A_1, A_2) &= \max_{a_1 \in A_1, a_2 \in A_2} \sum_{h_1 \in H_1, h_2 \in H_2} (U_1(a_1, h_1) + U_2(a_2, h_2))\, P(h_1, h_2) \\
&= \max_{a_1 \in A_1, a_2 \in A_2} \left(\sum_{h_1 \in H_1, h_2 \in H_2} U_1(a_1, h_1) P(h_1, h_2) \right. \\
&\quad \left. + \sum_{h_1 \in H_1, h_2 \in H_2} U_2(a_2, h_2) P(h_1, h_2) \right) \\
&= \max_{a_1 \in A_1} \sum_{h_1 \in H_1, h_2 \in H_2} U_1(a_1, h_1) P(h_1, h_2) \\
&\quad + \max_{a_2 \in A_2} \sum_{h_1 \in H_1, h_2 \in H_2} U_2(a_2, h_2) P(h_1, h_2) \\
&= \max_{a_1 \in A_1} \sum_{h_1 \in H_1} U_1(a_1, h_1) P(h_1) + \max_{a_2 \in A_2} \sum_{h_2 \in H_2} U_2(a_2, h_2) P(h_2) \\
&= MEU(A_1) + MEU(A_2).
\end{aligned}
$$

The proof of (ii) is similar.

Note

- The theorem is easily extended to any number of sets of nonintervening actions.

- The calculation trick from Section 6.2 can be applied to solve the two local decision problems.

6.3 Intervening actions

In this section we shall deal with intervening actions. That is, actions which have an impact on variables in the network which in turn affect the belief in the hypothesis variable (see Fig. 6.8). The actions could be treatments of crop fields, the hypothesis variable can be the state of the crop at the time of harvesting and the utility can be the value of the harvest.

6.3.1 One set of intervening actions

We have an action variable A, a utility function $U(H)$ over the hypothesis variable H (see Fig. 6.8), and suppose we have the evidence e.

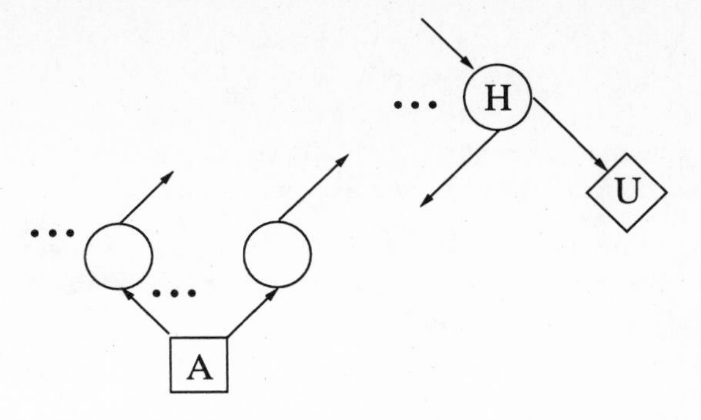

Figure 6.8 One set of intervening actions.

Then

$$EU(a \mid e) = \sum_{h \in H} U(h)P(h \mid a, e)$$

and

$$\text{Opt}(A \mid e) = \arg\max_{a \in A} EU(a \mid e).$$

A straightforward way of solving the decision problem is to simulate each possible action, a, to calculate $P(H \mid a)$ and then calculate the expected utility.

However, the trick from Section 6.2 can also be used in a modified form. If A is added to the network with even prior probability, and if U is added with $P(U = y \mid H) = NU(H)$ then, as in Section 6.2,

$$ENU(a \mid e) = P(U = y \mid a, e)$$

$$\text{Opt}(A \mid e) = \arg\max ENU(a \mid e).$$

So, if there are n different actions the optimal decision can be read from the network through n propagations.

If the network is large and the propagations costly, there is a more efficient way closely related to the one from Section 6.2. Let us see what happens if we enter $U = y$ as evidence and propagate. Then

$$\begin{aligned}
\text{Opt}(A \mid e) &= \arg\max P(U = y \mid a, e) \\
&= \arg\max \left(\frac{P(a \mid U = y, e) P(U = y \mid e)}{P(a \mid e)} \right) \\
&= \arg\max \left(\frac{P(a \mid U = y, e)}{P(a \mid e)} \right).
\end{aligned}$$

Unfortunately, $P(A \mid e)$ need not be an even distribution. However, $P(A \mid e)$ is known before $U = y$ is entered, and *the optimal action is the one with the largest relative increase when* $U = y$ *is entered.*

Note. If no descendent of A has received evidence, then A is d-separated from all variables with evidence entered. In that case we have $P(A \mid e) = P(A)$, and the optimal action can be read directly from $P(A \mid U = y, e)$.

This trick can be used when the cost of all actions are the same. If this is not the case we have to subtract costs from the expected utility.

Mildew. Two months before the harvest of a wheat field we observe the state Q of the crop and we observe whether it has been attacked by mildew, M. If there is an attack we should decide on a treatment with fungicides.

There are five variables.

- Q with states fair (f), average (a), good (g) and very good (v).
- M with states *no*, little (l), moderate (m) and severe (s).
- H (state of the crop at time of harvest) with states rotten (r), bad (b), poor (p) and the ones from Q. (Farmers in all countries tend to describe their harvest in pessimistic terms.)
- OQ (observation of Q) with the same states as Q.
- OM (observation of M) with the same states as M.

Furthermore, there is an action node A with actions *no*, light (l), moderate (m) and heavy (h), and a variable M^* describing the mildew attack after the action.

Both H and A have utility tables attached, and the total utility is the sum of the two ($C(A) \leq 0$).

Figure 6.9 gives a model.

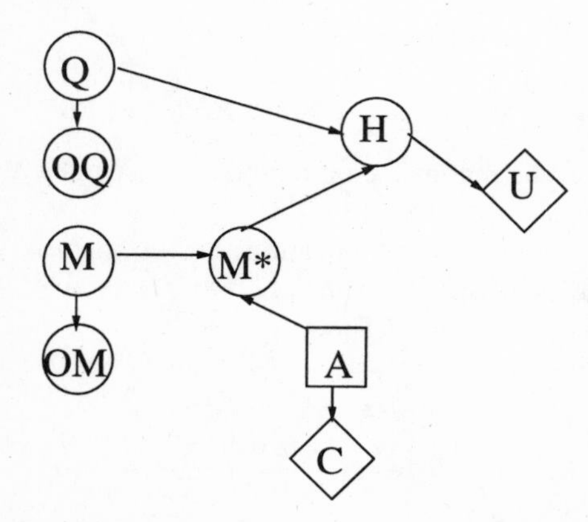

Figure 6.9 A decision model for Mildew.

By use of the normal trick on U we can, in one propagation, get $ENU(A)$ as

$$ENU(A \mid e) = \frac{P(A \mid U = y, e)P(U = y \mid e)}{P(A \mid e)}.$$

Then

$$EU(A \mid e) = \frac{1}{\alpha}(ENU(A \mid e) - \beta)$$

and it is a straightforward matter to take costs into account:

$$\text{Opt}_1(A \mid e) = \arg\max_{a \in A}(EU(a \mid e) + C(a)).$$

6.3.2 Several sets of intervening actions

Now, assume that we have two sets A_1 and A_2 of intervening actions, and a utility function $U(H)$ over the hypothesis variable H (see Fig. 6.10), and suppose we have the evidence e. (In fact, there may be several hypothesis variables, but this does not change the analysis.)

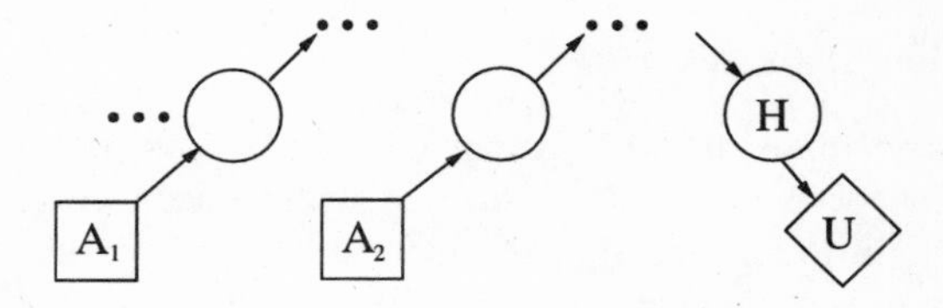

Figure 6.10 Two sets of intervening actions.

Then

$$EU(a_1, a_2 \mid e) = \sum_{h \in H} U(h)P(h \mid a_1, a_2, e)$$

and

$$\text{Opt}(A_1, A_2 \mid e) = \arg \max_{a_1 \in A_1, a_2 \in A_2} EU(a_1, a_2 \mid e).$$

Again, a straightforward way of solving the decision problem is to simulate each possible pair (a_1, a_2) of actions in order to calculate $P(H \mid a_1, a_2, e)$. However, this method becomes intractable when the number of action variables grows, and you would look for methods which decrease the number of propagations.

The calculation trick of using normalized utilities and entering $U = y$ yields (see Exercise 6.8):

$$\text{Opt}(A_1, A_2 \mid e) = \arg \max_{a_1 \in A_1, a_2 \in A_2} \frac{P(a_1, a_2 \mid U = y, e)}{P(a_1, a_2 \mid e)}.$$

Now, even if A_1 and A_2 are independent given e, and even if no descendents of A_1 and A_2 have received evidence, we are still left with a request for $P(A_1, A_2 \mid U = y, e)$, and there is not much hope that A_1 and A_2 are independent given $U = y$. Variable propagation can be used here.

6.4 Sequences of actions

In Sections 6.2.1 and 6.3.2 about several sets of actions, we dealt with a situation where we had to decide on a simultaneous tuple of actions. This does not mean that – for example – two buttons actually have to be pressed simultaneously. It means that in the time interval between the two actions you do not get any information that you did not have access to already.

Very often you are in a situation where you have to decide on an action A now, knowing that later in time, when more evidence on the system is acquired, you have to take a decision on actions in B. To analyze your present decision problem on A you have to imagine what you will do when deciding on B, and you have to imagine all possible information scenarios.

A thorough analysis of decision scenarios with sequences of actions is rather complicated, and it is out of the scope of this book. However, to give a hint of the problems involved we shall give several examples.

6.4.1 Information and actions

Before analyzing examples of decision sequences we shall look a little into what information means for the analysis. When analyzing decision situations one has to be very careful in describing the information at hand for the decision maker. In particular it is important to distinguish between what **you** know and what the **decision maker** knows. To illustrate this, consider the following scenarios.

Scenario 1. A player *PL* is playing the simplified poker game from Section 3.1.2. By watching the opponent's change of cards he has a distribution $P(OH2)$ of the opponent's hand. He can now decide between giving up (and pay one dollar) or call the opponent (which will either make him loose or win two dollars, depending on whose hand is best). A decision model may look like the one in Figure 6.11.

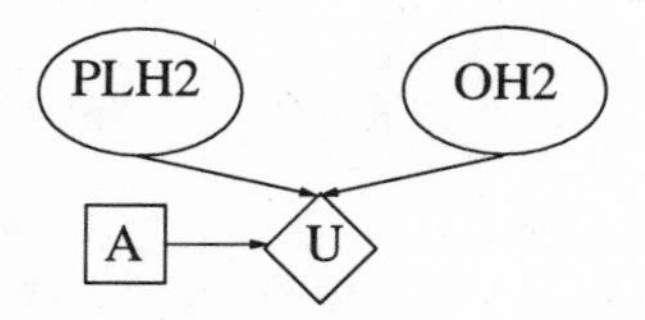

Figure 6.11 A model of Scenario 1.

For each type of hand $t \in PLH2$ the player will perform the following calculation for each action a:

$$EU(a \mid t) = \sum_{h \in OH2} U(a, t, h) P(h)$$

and

$$MEU(A \mid t) = \max_{a \in A} EU(a \mid t) \tag{6.2}$$

$$\text{Opt}(A \mid t) = \arg\max_{a \in A} EU(a \mid t).$$

Scenario 2. Suppose you are sitting at another table watching the two players. You see their change of cards but you cannot see their hands. What would you expect the player *PL* to win?

In contrast to you, the player knows his hand and he will act according to scenario 1. Then your expectation will be the mean of (6.2) over *PLH2*:

$$MEU(A) = \sum_{t \in PLH2} P(t) \max_{a \in A} \sum_{h \in OH2} U(a, t, h,)P(h).$$

Scenario 3. The player asks what you would do, without showing you his hand. Then

$$EU(a) = \sum_{t \in PLH2} P(t) \sum_{h \in OH2} P(h)U(a, t, h)$$

and

$$MEU(A) = \max_{a \in A} \sum_{t \in PLH2} \sum_{h \in OH2} P(t)P(h)U(a, t, h).$$

The difference between scenario 2 and scenario 3 is that in scenario 2 **you** only have probability distributions on the two hands but **the decision maker** knows the state of one of the hands, while in scenario 3 the decision maker (you) does not know any of the hands. Mathematically it has the impact that a summation operation and maximum operation are interchanged.

This can be formulated in the following *Sum-max-sum Rule*.

If A is an action variable, B a variable which has a state that is known by the decision maker, and C a variable which has a state that is not known, then

$$MEU(A) = \sum_{B} \max_{A} \sum_{C} U(A, B, C)P(B, C \mid A) \tag{6.3}$$

(B and C may be several variables).

In words:

> *First average over unknown variables, then maximize over the actions, and finally average over the variables known by the decision maker, but not known by you, the analyst.*

In graphical representations of decision problems, a link from a variable V to an action variable A indicates that the state of V is known when deciding on A. Such representations are called *influence diagrams.* So, Figure 6.11 is an influence diagram for scenario 3. An influence diagram for scenario 2 is given in Figure 6.12.

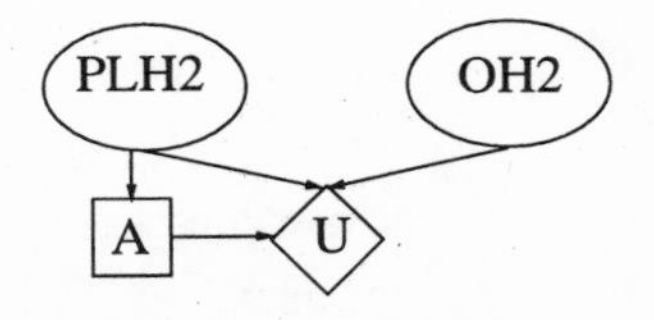

Figure 6.12 An influence diagram for Scenario 2.

6.4.2 Mildew examples

Extend the mildew example in Section 6.3.1 with an option T of deciding the time of harvesting; this decision is taken two months after the decision on the use of fungicides.

The approach for solving the decision problem for A is to assume that two months later, the decision on T will be optimal. So, for each action a we shall calculate $MEU(T, a)$, the expected utility provided T is decided optimally. Then $MEU(A) = \max_a MEU(T, a)$. This means that we – at the time of deciding on A – shall consider the decision problem at the time of deciding on T and imagine how the decision will be taken.

A crucial factor for calculating $MEU(T, a)$ is: *What kind of information is available at the time of deciding on* T?

Scenario 1 – No further information. The influence diagram in Figure 6.13 shows

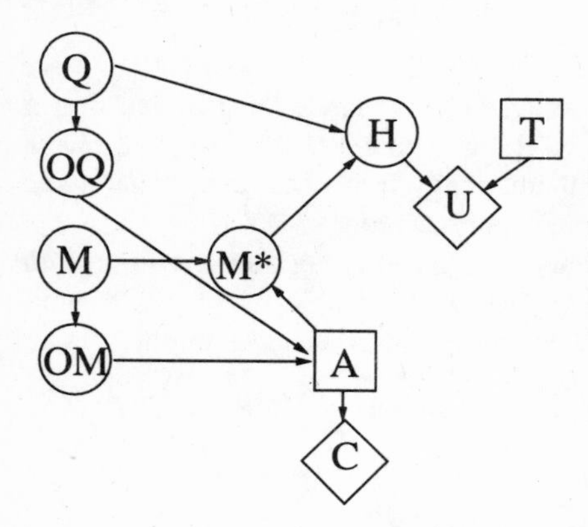

Figure 6.13 Influence diagram for Scenario 1. No further information is acquired between deciding on A and deciding on T.

that at the time of decision on T only a probability distribution on h can be used. If action a is taken it is $P(H \mid a)$ (we skip the conditioning over OQ and OM).

The maximal utility is then

$$MEU_1(T, a) = \max_t \left(\sum_h U(h, t) P(h \mid a) \right).$$

See also Section 6.2.1.

Scenario 2 – The state of H is known. In this scenario $P(H \mid a)$ is used to anticipate what the information on H actually will be. If it is h then the utility will be the maximum over T of $U(T, h)$. The probability for this is $P(h \mid a)$, so

$$MEU_2(T, a) = \sum_h \max_t U(t, h) P(h \mid a).$$

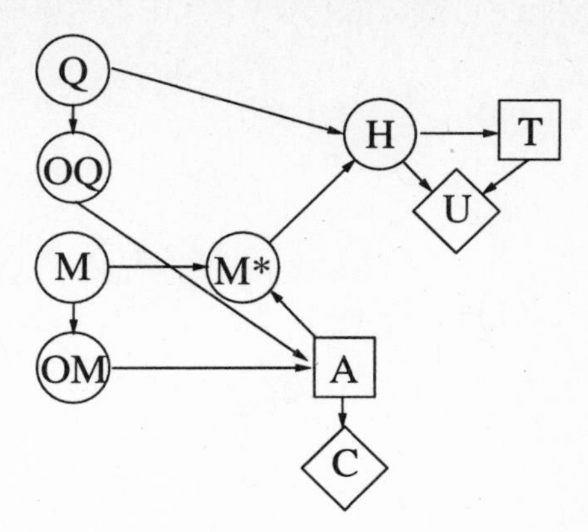

Figure 6.14 Influence diagram for Scenario 2. The state of H is known when deciding on T.

Scenario 3 – An observation on H is available. In this scenario new information on H will be available, but it is imprecise (see Fig. 6.15).

Now, $P(H \mid a)$ is not enough for the calculation of expected utilities. We also have to take into account that when deciding T only imperfect knowledge of H is available. If the observation of H is o then the maximal expected utility is (according to the sum-max-sum rule 6.3):

$$MEU_3(T, a \mid o) = \max_t \left(\sum_h U(t, h) P(h \mid a, o) \right).$$

However, at the time of deciding A we do not know the observation o. The best we have is $P(OH \mid a)$. Therefore

$$\begin{aligned} MEU_3(T, a) &= \sum_o P(o \mid a) \max_t \left(\sum_h U(t, h) P(h \mid a, o) \right). \\ &= \sum_o \max_t \left(\sum_h U(t, h) P(o, h \mid a) \right). \end{aligned}$$

Note. $MEU_1(a) \leq MEU_3(a) \leq MEU_2(a)$ (See Exercise 6.7), so the maximal expected utility increases with the quality of the information available.

As may be seen from the three scenarios above, decision analysis for sets of actions involve consideration of the sequence in which decisions are taken and the information available at the time of decision. In general it is possible to perform an analysis by simulating all action sequences together with all information configurations. However, this analysis grows exponentially with the sets of actions and the attached information variables.

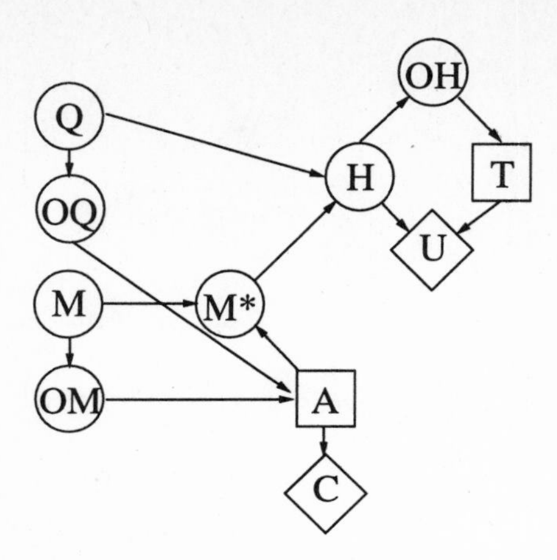

Figure 6.15 Influence diagram for Scenario 3. An observation on $H(OH)$ is acquired before deciding on T.

6.5 Summary

One set of nonintervening actions

A is a set of mutually excluding action options, H is a determining variable (may be a product of several variables). No action has an impact on $P(H)$. There is a utility table $U(A, H)$. Evidence e is entered.

The expected utility from taking action a is

$$EU(a \mid e) = \sum_{H} U(A, H)P(H \mid e).$$

The optimal action is $\text{Opt}(A \mid e) = \arg\max_A EU(a \mid e)$.

The maximal expected utility is $MEU(A \mid e) = \max_A EU(A \mid e)$.

Simulation in Bayesian networks

- Perform a linear transformation of $U(A, H)$ such that the minimal utility is zero, and the maximal utility is one. The normalized utility is

$$NU(A, H) = \alpha U(A, H) + \beta,$$

  where $\beta = -\min U(A, H)$ and $\alpha^{-1} = \max(U(A, H) + \beta)$.

- Add two variables A (with the actions as states) and U (with states y and n).

- Let U be a child of A and H.

- Give A the even prior distribution.

- Give $P(U = y \mid A, H)$ the table $NU(A, H)$.

- Enter $U = y$ as evidence and propagate.

Then the probabilities of A are proportional to the expected normalized utilities:

$$P(A \mid U = y, e) = kENU(A \mid e).$$

Several sets of nonintervening actions

$A_1, \ldots, A_n$ are action variables with determining variables $H_1, \ldots, H_n$. If $U(A_1, \ldots, A_n, H_1, \ldots, H_n) = \sum_i U_i(A_i, H_i)$, then:

- $MEU(A_1, \ldots, A_n) = \sum_i MEU(A_i)$;
- $\text{Opt}(A_1, \ldots, A_n) = (\text{Opt}(A_1), \ldots, \text{Opt}(A_n))$.

One set of intervening actions

Following the simulation procedure from nonintervening actions we get

$$\text{Opt}(A \mid e) = \arg\max_A \left(\frac{P(A \mid U = y, e)}{P(A \mid e)} \right)$$

and an optimal action is one with largest relative increase when $U = y$ is entered.

Several sets of intervening actions

The simulation procedure yields

$$\text{Opt}(A_1, \ldots, A_n) = \arg \max_{A_1, \ldots, A_n} \frac{P(A_1, \ldots, A_n \mid U = y, e)}{P(A_1, \ldots, A_n \mid e)}.$$

To get the joint probabilities for $A_1, \ldots, A_n$ variable propagation can be used.

Sequences of actions

$A_1, \ldots, A_n$ is a sequence of sets of actions to be decided in that order of time. $U(A_1, \ldots, A_n, H_1, \ldots, H_n)$ is the overall utility function. The set U is the universe of chance variables. $V_1, \ldots, V_{n+1}$ is a partition of the chance variables such that V_i is the set of variables instantiated before deciding on A_i (V_{n+1} is never instantiated.) Then

$$MEU(A_1, \ldots, A_n) = \sum_{V_1} \max_{A_1} \sum_{V_2} \ldots \sum_{V_n} \max_{A_n} \sum_{V_{n+1}} \varphi,$$

where $\varphi = U(A_1, \ldots, A_n, H_1, \ldots H_n) P(U \mid V_1, \ldots, V_n, A_1, \ldots A_n)$.

Table 6.7 Tables for Exercise 6.3.

	Carrier	Pure		Carrier	Pure
Out	−10	−10	Out	−3	−3
In	−40	100	In	−10	40
	Stallions			Mares	

6.6 Bibliographical notes

Decision theory has a long history, but achieved a breakthrough in the work of von Neumann & Morgenstein (1953) The structure they used was *decision trees*. Influence diagrams were proposed as an alternative to decision trees in Howard & Matheson (1984). The trick for using Bayesian networks to solve single decision problems is due to Cooper (1988). Methods for solving multiple decision problems through influence diagrams are constructed by Shachter (1986) and Shenoy (1992), and Jensen et al. (1994) use HUGIN propagation for solving influence diagrams.

Exercises

Exercise 6.1 Consider the example *management of effort* in Section 6.1.2.

(i) Let the marks be 0, 5, 6, 8, 9, 10. What is the optimal decision if the numerical values are used as utilities?

(ii) Consider the approach where the marks are given subjective utilities. Show that action *Gd* can only be optimal if the mark 0 is given higher utility than mark 3.

Exercise 6.2 ^{H} Extend Exercise 3.10 with the following.

In golf, the task is to use as few strokes as possible at each hole. I am driving at a 260 m long hole. If the drive is 265 m, I shall, on average, use 1.8 strokes to finish the hole. If the drive is 240 m, on average 2 extra strokes are needed; 220 m requires 2.5 extra strokes; 200 m requires 2.7; 180 m 2.9 extra strokes; 160 m 3.1; 145 m 3.3; a drive of 290 m will carry the ball out in hazards requiring 3.5 extra strokes; and if the drive is a miss, the ball will drop into a lake and it will require 4.5 extra strokes to finish the hole.

Construct a system which helps me decide between spoon and driver in the drive.

Exercise 6.3 ^{H} Consider the stud farm example from Section 3.2.1. Extend the model to be an aid for deciding for each horse whether it should be taken out of breeding or not. Table 6.7 gives the utilities.

Exercise 6.4 Consider the decision problem represented by the influence diagram below.

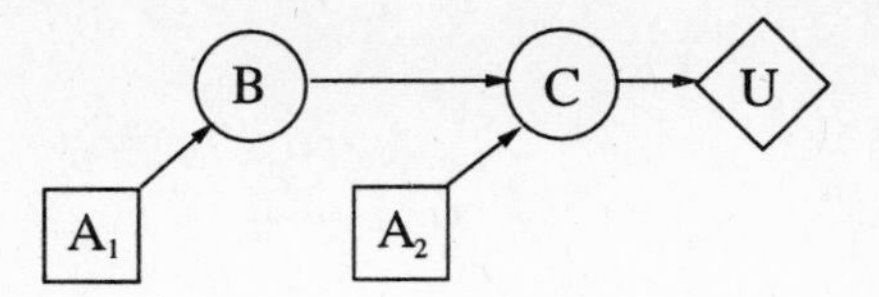

Figure for Exercise 6.4.

Show that the maximal expected utility for A_1, A_2 is

$$MEU(A_1, A_2) = \max_{A_1, A_2} \sum_B \sum_C U(C) P(B \mid A_1) P(C \mid A_2, B).$$

Exercise 6.5 Consider the decision problem represented by the influence diagram below.

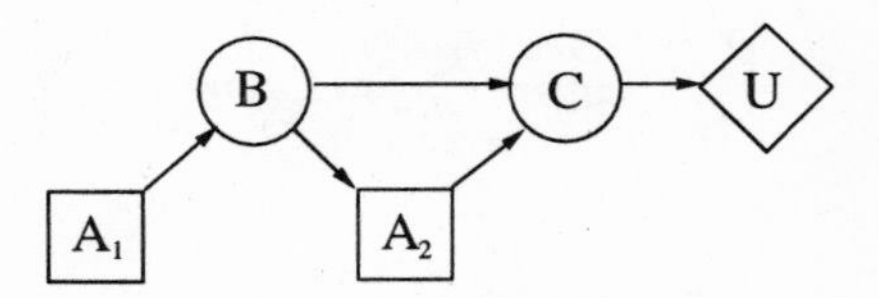

Figure for Exercise 6.5.

Show that the maximal expected utility for A_1 is

$$MEU(A_1) = \max_{A_1} \sum_B P(B \mid A_1) \max_{A_2} \sum_C U(C) P(C \mid A_2, B).$$

Compare the result with Exercise 6.4.

Exercise 6.6 ^{H} The mildew example (Section 6.3.1) has Table 6.8 and $U(H) = (-1, 1, 5, 8, 10, 12, 13)$, $C(A) = (0, -2, -3, -4)$, $P(Q) = (0.2, 0.4, 0.3, 0.1)$, $P(M) = (0.4, 0.3, 0.2, 0.1)$. Construct the system.

Exercise 6.7 (i) Let $\{a_{ij}\}$ be an $n \times m$ matrix of reals. Prove that

$$\max_i \sum_j a_{ij} \leq \sum_j \max_i a_{ij}.$$

(ii) Let A, B and C be variables, and let $U(A, B)$ be a a real function of the states of A and B. Prove that

$$\max_a \sum_C \sum_B U(a, B) P(C) P(B \mid C) = \max_a \sum_B U(a, B) P(B).$$

(iii) Use (i) and (ii) to prove for MEU_1, MEU_2 and MEU_3 from Section 6.4 that

$$MEU_1(a) \leq MEU_3(a) \leq MEU_2(a).$$

Exercise 6.8 Consider the situation in Section 6.3.2.

(i) Show that

$$P(a_1, a_2 \mid U = y, e) = \sum_h \frac{P(U = y \mid h)P(h, a_1, a_2 \mid e)}{P(U = y \mid e)}.$$

(Hint. U is independent of A_1, A_2 given H (Bayes' rule).)

(ii) Show that

$$P(a_1, a_2 \mid U = y, e) = \frac{P(a_1, a_2 \mid e)}{P(U = y \mid e)} \sum_h P(U = y \mid h)P(h \mid a_1, a_2, e).$$

(Hint. $P(H, a_1, a_2 \mid e) = P(H \mid a_1, a_2, e)P(a_1, a_2 \mid e)$.)

(iii) Let $P(U = y \mid H) = \alpha U(H) + \beta$. Show that

$$\arg \max_{a_1 \in A_1, a_2 \in A_2} \frac{P(a_1, a_2 \mid U = y, e)}{P(a_1, a_2 \mid e)} = \arg \max_{a_1 \in A_1, a_2 \in A_2} \sum_h U(h)P(h \mid a_1, a_2, e).$$

(iv) Show that

$$\text{Opt}(A_1, A_2 \mid e) = \arg \max_{a_1 \in A_1, a_2 \in A_2} \frac{P(a_1, a_2 \mid U = y, e)}{P(a_1, a_2 \mid e)}.$$

Exercise 6.9 ^{H} Consider the simplified poker game from Chapter 3. Let the stakes be so that each player has paid one dollar for entrance (and change of cards). The next decision is whether to call (costs one dollar) or to give up. Solve the following decision problems.

(i) You are the first to decide. If you call, your opponent is forced to follow (and pay a dollar), and the one with the best hand collects all four dollars. If you give up, your opponent collects the two dollars for entrance.

(ii) The second player can choose between following and giving up. You are the second player. (**Note.** You get information not only from your opponents change of cards but also from his decision whether to call or not.)

(iii) As (ii), however, here you are the first to decide.

(iv) In the light of (iii), can the infinite regression be solved?

Table 6.8 Tables for Exercise 6.6.

	f	a	g	v
f	0.8	0.3	0.1	0
a	0.15	0.6	0.2	0.1
g	0.05	0.1	0.6	0.4
v	0	0	0.1	0.5

$P(OQ \mid Q)$

	no	l	m	s
no	0.9	0.2	0.1	0
l	0.1	0.5	0.2	0.1
m	0	0.2	0.5	0.3
s	0	0.1	0.2	0.6

$P(OM \mid M)$

	f	a
no	(0, 0.05, 0.1, 0.7, 0.1, 0.05, 0)	(0, 0, 0.05, 0.1, 0.7, 0.1, 0.05)
l	(0.05, 0.1, 0.7, 0.1, 0.05, 0, 0)	(0, 0.05, 0.1, 0.7, 0.1, 0.05, 0)
m	(0.15, 0.7, 0.1, 0.05, 0, 0, 0)	(0.05, 0.1, 0.7, 0.1, 0.05, 0, 0)
s	(0.9, 0.1, 0, 0, 0, 0, 0)	(0.15, 0.7, 0.1, 0.05, 0, 0, 0)

	g	v
no	(0, 0, 0, 0.05, 0.1, 0.7, 0.15)	(0, 0, 0, 0, 0.1, 0.2, 0.7)
l	(0, 0, 0.05, 0.1, 0.7, 0.15, 0)	(0, 0, 0, 0.05, 0.15, 0.7, 0.1)
m	(0, 0.05, 0.1, 0.7, 0.1, 0.05, 0)	(0, 0, 0.05, 0.1, 0.7, 0.15, 0)
s	(0.05, 0.1, 0.7, 0.1, 0.05, 0, 0)	(0, 0.05, 0.1, 0.7, 0.1, 0.05, 0)

$P(H \mid Q, M^*)$

	no	l	m	h
no	(1, 0, 0, 0)	(1, 0, 0, 0)	(1, 0, 0, 0)	(1, 0, 0, 0)
l	(0, 1, 0, 0)	(0.8, 0.2, 0, 0)	(1, 0, 0, 0)	(1, 0, 0, 0)
m	(0, 0, 1, 0)	(0, 0.8, 0.2, 0)	(0.8, 0.2, 0, 0)	(1, 0, 0, 0)
s	(0, 0, 0, 1)	(0, 0, 0.8, 0.2)	(0, 0.8, 0.2, 0)	(0.8, 0.2, 0, 0)

$P(M^* \mid A, M)$

Appendix A

Construction of junction trees (proofs)

This appendix contains proofs of the crucial theorems in Section 4.5.

Definitions. Let G be an undirected graph with node set N.

If in G there is a link between A and B they are said to be *neighbours*. Sometimes the word *adjacent* is used. A *path* in G is a sequence $A_1, \ldots, A_n$ of distinct nodes where A_i and A_{i+1} are neighbours. A *cycle* is a path where $A_1 = A_n$ and all other nodes are distinct. A *chord* in a cycle $A_1, \ldots, A_n$ is a link between two nodes A_i and A_j, where i and j are not consecutive numbers.

G is *triangulated* if any cycle of length > 3 has a chord.

A subset S of N is *complete* if each pair of nodes in S are neighbours. A node is *simplicial* if its neighbour set is complete.

A node A is *eliminated* from G by adding links to G such that A becomes simplicial and then removing it together with its links. The result is denoted G_A. G is said to be *eliminatable* if all nodes can be successively eliminated without adding extra links.

Theorem A.1 *Any eliminatable graph is triangulated.*

Proof. Induction on the number of nodes.

Clearly, a graph consisting of one node is both triangulated and eliminatable.

Assume the theorem to hold for all graphs consisting of $< n$ nodes, and let G be an eliminatable graph with n nodes. Since G is eliminatable it must have at least one simplicial node A. By the induction hypothesis, G_A is triangulated. Since A is simplicial, any cycle of length > 3 containing A must have a chord (see Fig. A.1). Hence G is triangulated.

To prove that any triangulated graph is eliminatable we need a lemma.

Lemma A.1 (Decomposition lemma.) *Let G be a noncomplete triangulated graph with at least three nodes, and with node set N. Then there is a complete subset S of N such that $G \setminus S$ is disconnected.*

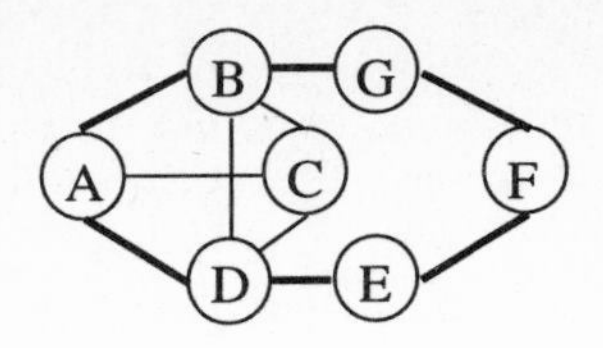

Figure A.1 A cycle containing a simplicial node must have a chord.

Proof. Let A and B be two nonadjacent nodes, and let S be a minimal set of nodes such that any path between A and B contains a node from S. Clearly $G \setminus S$ is disconnected, and let G^A denote the connected component containing A joined with S (see Fig. A.2). G^B is defined similarly.

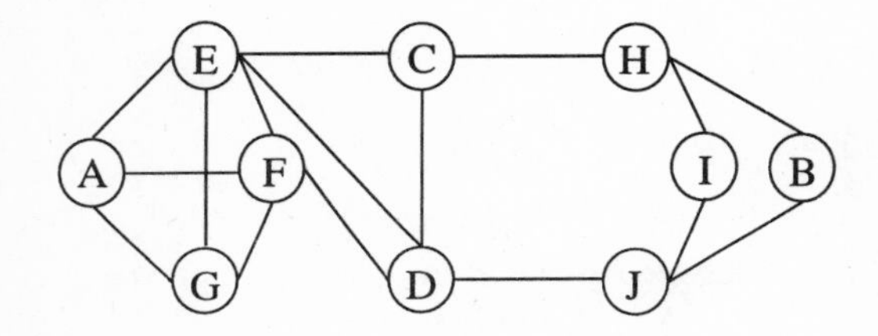

Figure A.2 A triangulated graph with a separator consisting of $\{C, D\}$. G^A consists of the nodes A, C, D, E, F, and G.

We only need to prove that S is complete. Let C and D be nodes from S. We shall prove that they are neighbours.

Since S is minimal there must be a path from A to B with C as the only node from S; the same holds for D. Now, take a shortest path in G_A connecting C and D. Do the same for G_B and join the two paths in a cycle of length > 3. This cycle must have a chord, and the only chord possible is a link between C and D.

Theorem A.2 *Any triangulated graph is eliminatable.*

Proof. Induction on the number of nodes.

Clearly, any graph with at most two nodes is both triangulated and eliminatable. To carry out an induction proof it would suffice to prove that any triangulated graph contains at least one simplicial node. If this is the case, then a triangulated graph with n nodes is reduced by eliminating a simplicial node, and the induction hypothesis will yield the graph eliminatable. For technical reasons we shall use a slightly stronger result.

Claim. Let G be a triangulated graph containing at least two nodes. Then it contains at least two simplicial nodes. If G is not complete then the two nodes can be chosen nonadjacent.

As indicated above, the claim is sufficient to prove the theorem.

Proof of the claim. The claim is trivially true for any graph containing two nodes.

Now, assume the claim to be true for any graph with $< n$ nodes and let G have n nodes. If G is complete then any two nodes are simplicial. So, suppose that G is not complete. Then the decomposition lemma yields a complete subset S separating G into at least two components.

Let G_A and G_B be defined as in the proof of the decomposition lemma. They are both triangulated (any chordless cycle in G_A is also a chordless cycle in G).

If G_A is not complete we can choose two nonadjacent simplicial nodes A_1 and A_2. Since they cannot both be members of S, we choose A_1 outside S to ensure that A_1 is simplicial in G. If G_A is complete, choose any node outside S. The same can be done for G_B. Since both nodes are chosen outside S they are nonadjacent.

Let us recall the following.

Definitions. A *junction graph* for an undirected graph G is an undirected, labelled graph. The nodes are the cliques of G, and every pair of nodes with a nonempty intersection has a link labelled by that intersection.

A spanning tree of Gs junction graph is a *junction tree* for G if it has the property that for each pair U, V of nodes, all the nodes in the path between U and V contain the intersection $U \cap V$. The labels on the links in a junction tree are called *separators*.

Theorem A.3 *A connected undirected graph G is triangulated if it has a junction tree.*

Proof. Induction on the number of nodes in the graph.

The theorem is trivially true for any graph with at most two nodes.

Assume the theorem to hold for all graphs with $< n$ nodes. Let G be a connected graph with n nodes, and let T be a junction tree, for G. Since T is a tree there is a clique C with only one neighbour C' in T. Let $A \in C \setminus C'$. Since T is a junction tree, A can only be a member of one clique, namely C. Then all neighbours of A are members of C and hence pairwise linked. This means that A is simplicial in G.

Now, remove A from C (if thereby the new clique becomes a subset of C' then remove C from T). The resulting tree T^* is a junction tree for G_A.

By the induction hypothesis G_A is triangulated, and hence G is also triangulated.

Theorem A.4 *Any connected triangulated graph has a junction tree.*

Proof. Induction on the number of nodes of the graph.

Trivially, the theorem is true for any graph with at most two nodes.

Assume the theorem to hold for any graph with $< n$ nodes, and let G be a connected triangulated graph with n nodes. By the proof of Theorem A.2 G has at least one simplicial node A. Then A together with its neighbours form a clique C in G. Also, G and G_A have the same cliques except for C. G_A may instead have a clique consisting of $C \setminus \{A\}$. Also, G_A must be triangulated because G is, and – by the induction hypothesis – G_A has a junction tree T^*.

Now, construct T from T^* in the following way.

- If $S = C \setminus \{A\}$ is a clique in T^* then add A to S.

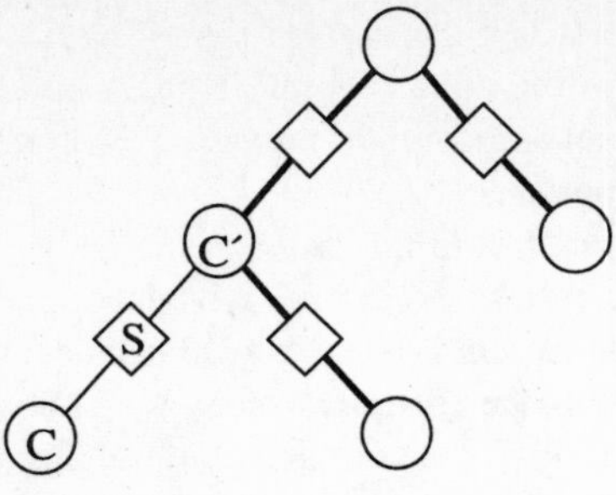

Figure A.3 The construction of a junction tree for G from a junction tree for G_A.

- If not, then S is a subset of a clique C' in T^*. In this case, add the clique C with a link to C' labelled S (see Fig. A.3).

It is easy to verify that T is a junction tree for G.

Appendix B

Value of information (proofs)

This appendix contains proofs omitted in Section 5.5.

Proposition B.1 *If* $V(P(H)) = \sum_{h \in H} a_h P(h)$ *then* $EV(T) = V(P(H))$.

Proof.

$$\begin{aligned} EV(T) &= \sum_{t \in T} P(t) V(P(H \mid t)) = \sum_{t \in T} P(t) \sum_{h \in H} a_h P(h \mid t) = \sum_{t \in T} \sum_{h \in H} a_h P(h, t) \\ &= \sum_{h \in H} \sum_{t \in T} a_h P(h, t) = \sum_{h \in H} a_h P(h) = V(P(H)). \end{aligned}$$

Proposition B.2 *(***Jensen's inequality***)*

(Jensen is the most frequent Danish name. This Jensen lived from 1859 to 1925.)

Let $V : R^n \to R$ *be a convex function. Let* $\underline{x}_1, \ldots, \underline{x}_n \in R^n$ *and let* $a_1, \ldots, a_n \in [0, 1]$*, such that* $\sum_{i=1}^{n} a_i = 1$*. Then*

$$V\left(\sum_{i=1}^{n} a_i \underline{x}_i\right) \leq \sum_{i=1}^{n} a_i V(\underline{x}_i).$$

Proof. Induction on n. The basis $n = 2$ is precisely the convexity property.

Assume that the inequality holds for $n - 1$.

Put $t = \sum_{i=1}^{n-1} a_i$, then $a_n = 1 - t$, and assume $t \neq 0$.

$$\begin{aligned} V\left(\sum_{i=1}^{n} a_i \underline{x}_i\right) &= V(t\left(\sum_{i=1}^{n-1} \frac{ai}{t} \underline{x}_i\right) + (1 - t)\underline{x}_n) \\ &\leq tV\left(\sum_{i=1}^{n-1} \frac{a_i}{t} \underline{x}_i\right) + (1 - t)V(\underline{x}_n) \\ &\leq t \sum_{i=1}^{n-1} \frac{a_i}{t} V(\underline{x}_i) + a_n V(\underline{x}_n) \end{aligned}$$

$$= \sum_{i=1}^{n} a_i V(\underline{x}_i).$$

Theorem B.1 *Let H be a variable with n states, let T be another discrete variable, and let $V : [0,1]^n \rightarrow R$ be a convex function.*

Then

$$\sum_{t \in T} V(P(H|t))P(t) \geq V(P(H)).$$

Proof. (Jensen's inequality)

$$\sum_{t \in T} V(P(H|t))P(t) \geq V\left(\sum_{t \in T} P(H \mid t)P(t)\right) = V\left(\sum_{t \in T} P(H,t)\right) = V(P(H)).$$

Proposition B.3 *The functions:* **entropy, variance** *and* **utility** *are convex functions, and* WOE *is convex for $x > \frac{1}{2}$.*

Proof. Let H be a variable, and let $P(H)$ and $P'(H)$ be two distributions over H. We shall prove that for each $t \in [0,1]$,

$$tV(P(H)) + (1-t)V(P'(H)) \geq V(tP(H) + (1-t)P'(H)).$$

Entropy. First note that the function $x \log x$ is convex for $x > 0$ (the second derivative is positive). So for all $x, y > 0$,

$$tx \log x + (1-t)y \log y \geq (tx + (1-t)y) \log(tx + (1-t)y).$$

Then

$$\begin{aligned}
& t(-ENT(P(H))) + (1-t)(-ENT(P'(H))) \\
&= \sum_{h \in H} [tP(h) \log P(h) + (1-t)P'(h) \log P'(h)] \\
&\geq \sum_{h \in H} (tP(h) + (1-t)P'(h)) \log(tP(h) + (1-t)P'(h)) \\
&= -ENT(tP(H) + (1-t)P'(H)).
\end{aligned}$$

Variance. $V(P(H)) = -\sum_{h \in H} (h - \mu)^2 P(h)$, where $\mu = \sum_{h \in H} hP(h)$. The constant μ is the one which minimizes $\sum_{h \in H} (h-k)^2 P(h)$.

We have

$$V(tP(H) + (1-t)P'(H)) = -\sum_{h \in H} (h - \mu_0)^2 (tP(h) + (1-t)P'(h)),$$

where $\mu_0 = \sum_{h \in H} h(tP(H)) + (1-t)P'(H)$. Then

$$\begin{aligned}
& tV(P(H)) + (1-t)V(P'(H)) \\
&= -t \sum_{h \in H} (h - \mu_1)^2 P(h) - (1-t) \sum_{h \in H} (h - \mu_2)^2 P'(H) \\
&\geq -t \sum_{h \in H} (h - \mu_0)^2 P(h) - (1-t) \sum_{h \in H} (h - \mu_0)^2 P'(h) \\
&= -\sum_{h \in H} (tP(h) + (1-t)P'(h))(h - \mu_0)^2 \\
&= V(tP(H) + (1-t)P'(H)).
\end{aligned}$$

Utility.

$$
\begin{aligned}
& V(tP(H) + (1-t)P'(H)) \\
&= \max_{a \in A} \sum_{h \in H} U(h,a)(tP(h) + (1-t)P'(h)) \\
&\leq \max_{a \in A} \sum_{h \in H} U(h,a)tP(h) + \max_{a \in A} \sum_{h \in H} U(h,a)(1-t)P'(h) \\
&= tV(P(H)) + (1-t)V(P'(H)).
\end{aligned}
$$

Weight of evidence.

$$WOE(x) = \log x - \log(1-x)$$

$$WOE''(x) = \frac{1}{(1-x)^2} - \frac{1}{x^2} > 0 \text{ for } x > \frac{1}{2}.$$

Theorem B.2 *Let H be a hypothesis variable, and let A be an action variable; let $U: H \times A \to R$ be the utility function. Define*

$$V(P(H)) = \max_{a \in A} \sum_{h \in H} P(h)U(h,a).$$

Then $EV(T) = V(P(H))$ if and only if no outcome of T will change the optimal action of A.

Proof. Let the optimal action be a_0, then

$$V(P(H)) = \sum_{h \in H} P(h)U(h,a_0).$$

If the maximal action is not changed no matter the outcome of T, then

$$EV(T) = \sum_{t \in T} P(t) \sum_{h \in H} P(h \mid t)U(h,a_0) = \sum_{h \in H} P(h)U(h,a_o) = V(P(H)).$$

Now, suppose that $EV(T) = V(P(H))$, then

$$\sum_{t \in T} \left(\max_{a \in A} \left(\sum_{h \in H} P(h,t)U(h,a) \right) - \sum_{h \in H} P(h,t)U(h,a_0) \right) = 0.$$

Since

$$\max_{a \in A} \sum_{h \in H} P(h,t)U(h,a) \geq \sum_{h \in H} P(h,t)U(h,a_0)$$

for all t, we conclude that equality holds for all t. This means that for each t, a_0 is an optimal action.

Bibliography

Abramson, B., J. Brown, W. Edwards, A. Murphy, R. Winkler 1996. Hailfinder: A Bayesian system for forecasting severe weather. *International Journal of Forecasting*, to appear.

Andreassen, S., R. Hovorka, J. Benn, K. Olesen, E. Carson 1991. A model-based approach to insulin adjustment. In *Proceedings of the third conference on artificial intelligence in medicine*, Stefanelli, Hasman, Fieschi, Tatma (eds), 239–48. Maastricht: Springer.

Andreassen, S., F. V. Jensen, S. K. Andersen, B. Falck, U. Kjærulff, M. Woldbye, A. Sørensen, A. Rosenfalck, F. Jensen 1989. MUNIN – an expert EMG assistant. In *Computer-aided electromyography and expert systems*, J. E. Desmedt (ed.), 255–77. Amsterdam: Elsevier Science.

Beeri, C., R. Fagin, D. Maier, M. Yannakakis 1983. On the desirability of acyclic database schemes. *Journal of the Association for Computing Mashinery* **30**(3), 479–513.

Ben-Bassat, M. 1978. Myopic policies in sequential classification. *IEEE Transactions of Computing* **27**, 170–4.

Bertele, U. & Brioschi 1972. *Nonserial dynamic programming*. London: Academic.

Binford, T., T. Levitt, W. Mann 1988. Bayesian inference in model-based machine vision. In *Uncertainty in artificial intelligence*, J. F. Lemmer & L. M. Kanal (eds). Amsterdam: Elsevier Science.

Bruza, P. & L. van der Gaag 1993. Efficient context-sensitive plausible inference for information disclosure. *Proceedings of the 16th international conference on research and development in information retrieval f(ACM-SIGIR '93)*, 12–21.

Buntine, W. L. 1994. Operations for learning with graphical models. *Journal of Artificial Intelligence Research* **2**, 159–225.

Burnell, L. & E. Horvitz 1995. Structure and chance: Melding logic and probability for software debugging. *Communications of the ACM* **38**(3), 31–41.

Cannings, C., E. Thompson, M. Skolnick 1978. Probability functions on complex pedigrees. *Advanced Applied Probability* **10**, 26–61.

Charniak, E. 1991. Bayesian networks without tears. *AI Magazine* **12**(4), 50–63.

Charniak, E. & R. Goldman 1991. Probabilistic abduction for plan-recognition. *Proceedings of the ninth national conference on artificial intelligence*, 160–65. American Association for Artificial Intelligence.

Cooper, G. F. 1988. A method for using belief networks as influence diagrams. In *Proceedings of the fourth workshop on uncertainty in artificial intelligence, Minneapolis*, 55–63.

Cooper, G. F. 1990. The computational complexity of probabilistic inference using Bayesian belief networks. *Artificial Intelligence* **42**(2–3), 393–405.

Cooper, G. F. & E. Herskovits 1992. A Bayesian method for the induction of probabilistic networks from data. *Machine Learning* **9**(4), 309–48.

Cowell, R. & A. Dawid 1992. Fast retraction of evidence in a probabilistic expert system. *Statistics and Computing* **2**, 37–40.

Dawid, A. 1992. Applications of a general propagation algorithm for probabilistic expert system. *Statistics and Computing* **2**, 25–36.

Dawid, A. P. & S. L. Lauritzen 1993. Hyper Markov laws in the statistical analysis of decomposable graphical models. *The Annals of Statistics* **21**(3), 1272–317.

de Dombal, F., D. Leaper, J. Staniland, A. McCann, J. Harrocks 1972. Computer-aided diagnosis of acute abdominal pain. *British Medical Journal* **2**, 9–13.

Edwards, D. & T. Havranek 1985. A fast procedure for model search in multidimensional contingency tables. *Biometrika* **72**, 339–51.

Franklin, R., D. Spiegelhalter, F. Macartney, K. Bull 1989. Combining clinical judgements and statistical data in expert systems: over the telephone management decisions for critical congenital heart disease in the first month of life. *International Journal of Clinical Monitoring and Computing* **6**, 157–66.

Franklin, R., D. Spiegelhalter, F. Macartney, K. Bull 1991. Evaluation of a diagnostic algorithm for heart disease in neonates. *British Medical Journal* **302**, 935–9.

Fung, R. & S. Crawford 1990. CONSTRUCTOR – a system for induction of probabilistic models. *Proceedings of the eighth national conference on artificial intelligence*, 762–79. American Association for Artificial Intelligence.

Fung, R. & B. D. Favero 1994. Backward simulation in Bayesian networks. In *Proceedings of the tenth conference on uncertainty in artificial intelligence*, R. L. de Mantaras & D. Poole (eds) 227–34. San Francisco, CA: Morgan Kaufmann.

Fung, R. & B. D. Favero 1995. Applying Bayesian networks to information retrieval. *Communications of the ACM* **38**(3), 42–8.

Geman, S. & D. Geman 1984. Stochastic relaxation, Gibbs distributions, and the Bayesian restoration of images. *IEEE Transactions on Pattern Analysis and Machine Intelligence* **6**(6), 721–41.

Geyer, C. J. 1992. Practical Markov Chain Monte Carlo (with discussion). *Statistical Science* **7**, 473–511.

Gilks, W., A. Thomas, D. Spiegelhalter 1994. A language and a program for complex Bayesian modelling. *The Statistician* **43**, 169–78.

Glasziou, P. & J. Hilden 1989. Test selection measures. *Medical Decision Making* **9**, 133–44.

Goldman, R. & E. Charniak 1993. A language for construction of belief networks. *IEEE Transactions on Pattern Analysis and Machine Intelligence* **15**(3), 196–208.

Golumbic, M. C. 1980. *Algorithmic graph theory and perfect graphs*. London: Academic.

Gorry, G. 1973. Computer-assisted clinical decision making. *Methods of Information in Medicine* **12**, 45–51.

Gorry, G. & G. Barnett 1968. Experience with a model of sequential diagnosis. *Computers and Biomedical Research* **1**, 490–507.

Habbema, J. D. F. 1976. Models diagnosis and detection of diseases. In *Decision making and medical care*, de Dombal et al. (eds), 399–411. Amsterdam: Elsevier Science.

Heckerman, D. 1993. Causal independence, knowledge acquisition, and inference. In *Proceedings of the ninth conference on uncertainty in artificial intelligence*, D. Heckerman & A. Mamdani (eds), 122–7. San Mateo, CA: Morgan Kaufmann.

Heckerman, D., J. Breese, K. Rommelse 1995. Decision-theoretic troubleshooting. *Communications of the ACM* **38**(3), 49–56.

Heckerman, D., D. Geiger, D. M. Chickering 1994. Learning Bayesian networks: the combination of knowledge and statistical data. In *Proceedings of the 10th conference on uncertainty in artificial intelligence*, R. L. de Mantaras & D. Poole (eds), 293–301. San Francisco, CA: Morgan Kaufmann.

Heckerman, D., E. Horvitz, B. Nathwani 1992. Towards normative experts systems: Part I. the Pathfinder project. *Methods of Information in Medicine* **31**, 90–105.

Heckerman, D. & B. Nathwani 1992a. An evaluation of the diagnostic accuracy of Pathfinder. *Computers and Biomedical Research* **25**, 56–74.

Heckerman, D. & B. Nathwani 1992b. Towards normative experts systems: Part II. probability-based representations for efficient knowledge acquisition and inference. *Methods of Information in Medicine* **31**, 106–16.

Hejlesen, O. K., S. Andreassen, S. K. Andersen 1993. Implementation of a learning procedure for multiple observations in a diabetes advisory system based on causal probabilistic networks. In *Proceedings of the fourth conference on artificial intelligence in medicine*, S. Andreassen, B. Engelbrecht, H. Wyat (eds), 63–74. Amsterdam: IOS.

Henrion, M. 1988. Propagating uncertainty in Bayesian networks by probabilistic logic sampling. In *Uncertainty in artificial intelligence 2*, J. F. Lemmer & L. M. Kanal (eds), 149–63. Amsterdam: Elsevier Science.

Højsgaard, S. & B. Thiesson 1995. BIFROST – Block recursive models Induced From Relevant knowledge, Observations, and Statistical Techniques. *Computational Statistics & Data Analysis* **19**, 155–75.

Horvitz, E. & B. Barry 1995. Display of information for time-critical decision making. In *Proceedings of the eleventh conference on uncertainty in artificial intelligence*, Besnard & Hanks (eds), 296–305. San Fransisco, CA: Morgan Kaufmann.

Howard, R. A. 1966. Information value theory. *IEEE Transactions on Systems Science and Cybernetics*, 22–6.

Howard, R. A. & J. E. Matheson 1984. Influence diagrams. In *Readings on the principles and applications of decision analysis*, R. A. Howard & J. E. Matheson (eds), Vol. 2, 719–62. Menlo Park, CA: Strategic Decisions Group.

Jensen, A. L. 1995a. A probabilistic model based decision support system for mildew management in winter wheat. *Technical Report DINA Report No. 39*. DINA, Research Centre Foulum, PB 29, 8830 Tjele, Denmark.

Jensen, C. S., A. Kong, U. Kjærulff 1995. Blocking-Gibbs sampling in very large probabilistic expert systems. *International Journal of Human-Computer Studies* **42**, 647–66.

Jensen, F., F. V. Jensen, S. L. Dittmer 1994. From influence diagrams to junction trees. In *Proceedings of the tenth conference on uncertainty in artificial intelligence*, R. L. de Mantaras & D. Poole (eds), 367–73. San Fransisco, CA: Morgan Kaufmann.

Jensen, F. V. 1995b. Cautious propagation in Baysian networks. In *Proceedings of the eleventh conference on uncertainty in artificial intelligence*, P. Besnard & S. Hanks (eds), 323–8. San Fransisco, CA: Morgan Kaufmann.

Jensen, F. V., B. Chamberlain, T. Nordahl, F. Jensen 1991. Analysis in HUGIN of data conflict. In *Uncertainty in Artificial Intelligence 6*, N.-H. Bonnisone et al. (eds), 519–28. Amsterdam: North-Holland.

Jensen, F. V., H. I. Christensen, J. Nielsen 1992. Bayesian methods for interpretation and control in multi-agent vision systems. In *Proceedings from applications of artificial intelligence X*, K. W. Bowyer (ed.), 536–48. Orlando, FL: SPIE – The International Society for Optical Engineering.

Jensen, F. V. & F. Jensen 1994. Optimal junction trees, In *Proceedings of the tenth conference on uncertainty in artificial intelligence*, R. L. de Mantaras & D. Poole (eds), 360–66. San Fransisco, CA: Morgan Kaufmann.

Jensen, F. V., K. G. Olesen, S. K. Andersen 1990. An algebra of Bayesian belief universes for knowledge-based systems. *Networks* **20**, 637–59.

Kim, J. H. & J. Pearl 1983. A computational model for causal and diagnostic reasoning in inference systems. *Proceedings of the eighth international joint conference on artificial intelligence*, 190–93. American Association for Artificial Intelligence.

Kim, Y. & M. Valtorta 1995. On the detection of conflicts in diagnostic Bayesian networks using abstraction. In *Proceedings of the eleventh conference on uncertainty in artificial intelligence*, P. Besnard & S. Hanks (eds), 362–7. San Fransisco, CA: Morgan Kaufmann.

Lauritzen, S. L. 1996. *Graphical Models*. Oxford: Oxford University Press, in press.

Lauritzen, S. L. & D. J. Spiegelhalter 1988. Local computations with probabilities on graphical structures and their application to expert systems (with discussion). *Journal of the Royal Statistical Society, Series B* **50**, 157–224.

Lauritzen, S. L., B. Thiesson, D. J. Spiegelhalter 1994. Diagnostic systems created by model selection methods: a case study. In *Selecting models from data: AI and statistics IV (Lecture Notes in Statistics 89)*, P. Cheeseman & R. Oldford (eds), 143–52. New York: Springer.

Levitt, T., T. Binford, G. Ettinger 1989. Utility-based control for computer vision, In *Uncertainty in artificial intelligence*, Shachter, Levitt, Kanal, Lemmer (eds), 407–21. Amsterdam: Elsevier Science.

Levitt, T., M. Hedgcock, J. Dye, S. Johnston, V. Shadle, D. Vosky 1993. Bayesian inference for model-based segmentation of computed radiographs of the hand. *Artificial Intelligence in Medicine* **5**(4), 365–87.

Lindley, D. V. 1971. *Making decisions*. New York: Wiley.

Madigan, D. & K. Mosurski 1993. Explanation in belief networks. *Technical report*, University of Washington, US and Trinity College, Dublin, Ireland.

McDermott, J. 1984. R1 re-visited: 4 years in the trenches. *The Artificial Intelligence Magazine* (Fall), 21–32.

Munck-Fairwood, R. 1992. Recognition of geometric primitives using logic program and probabilistic network reasoning methods. *Proceedings of SPIE: applications in AI, X; machine vision and robotics*, 589–600. Orlando, FL: SPIE.

Nathwani, B., D. Heckerman, E. Horvitz, T. Lincoln 1990. Integrated expert systems and videodiscs in surgical pathology. *Human Pathology* **21**(1), 11–17.

Pearl, J. 1982. Reverend Bayes on inference engines: a distributed hierarchical approach. *Proceedings of American Association for Artificial Inteligence National Conference on AI, Pittsburgh*, 133–6.

Pearl, J. 1986a. A constraint-propagation approach to probabilistic reasoning. In *Uncertainty in artificial intelligence*, L. M. Kanal & J. Lemmer (eds), 357–70. Amsterdam: North-Holland.

Pearl, J. 1986b. Fusion, propagation, and structuring in belief networks. *Artificial Intelligence* **29**(3), 241–88.

Pearl, J. 1988. *Probabilistic reasoning in intelligent systems: networks of plausible inference, (Series in representation and reasoning).* San Fransisco, CA: Morgan Kaufmann.

Rasmussen, L. K. 1995a. Bayesian network for blood typing and parentage verification of cattle. *Dina research report no. 38.* Aalborg University, Department of Mathematics and Computer Science.

Rasmussen, L. K. 1995b. BOBLO: an expert system based on Bayesian networks to blood group determination of cattle. *Research report 16.* Research Center Foulum, Denmark, PB 23, 8830 Tjele, Denmark.

Rimey, R. & C. Brown 1994. Control of selective perception using Bayes nets and decision theory. *International Journal of Computer Vision* **12**, 173–207. Special issue in Active Vision.

Shachter, R. D. 1986. Evaluating influence diagrams. *Operations Research* **34**(6), 871–82.

Shafer, G. & P. Shenoy 1990. Probability propagation. *Annals of Mathematics and Artificial Intelligence* **2**, 327–52.

Shenoy, P. P. 1992. Valuation-based systems for bayesian decision analysis. *Operations Research* **40**(3), 463–84.

Shortliffe, E. H. 1976. *Computer-based medical consultation: MYCIN*. Amsterdam: Elsevier Science.

Spiegelhalter, D. & R. Cowell 1992. Learning in probabilistic expert systems. In *Bayesian Statistics 4*, J. Bernardo, J. Berger, A. Dawid, A. Smith (eds), 447–65. Oxford: Clarendon.

Spiegelhalter, D. J. & R. P. Knill-Jones 1984. Statistical and knowledge-based approaches to clinical decision-support systems. *Journal of the Royal Statistical Society, Series A* **147**, 35–77.

Spiegelhalter, D. & S. L. Lauritzen 1990. Sequential updating of conditional probabilities on directed graphical structures. *Networks* **20**, 579–605.

Suermondt, H. J. 1992. *Explanation in Bayesian belief networks*. PhD thesis, Knowledge Systems Laboratory, Medical Computer Science, Stanford University, Stanford, California 94305. Report No. STAN-CS-92-1417.

Tarjan, R. E. & M. Yannakakis 1984. Simple linear-time algorithms to test chordality of graphs, test acyclicity of hypergraphs, and selectively reduce acyclic hypergraphs. *SIAM Journal of Computing* **13**(3), 566–79.

Verma, T. S. 1987. Causal networks: semantics and expressiveness. *Proceedings of the third workshop on uncertainty in artificial intelligence*, 352–359.

von Neumann, J. & O. Morgenstein 1953. *Theory of games and economic behavior*. New york: Wiley.

Winterfeldt, D. & W. Edwards 1986. *Decision analysis and behavioral research*. Cambridge: Cambridge University Press.

Xiang, Y., B. Pant, A. Eisen, M. Beddoes, & D. Poole 1993. Multiply sectioned Bayesian networks for neuromuscular diagnosis. *Artificial Intelligence in Medicine* **5**(4), 293–314.

Xu, H. 1994. Computing marginals from the marginal representation in Markov trees. *Proceedings from the international conference on information processing and management of uncertainty in knowledge-based systems (IPMU)*, 275–80. Cite Internationale Universitaire, Paris, France.

Index